The Dark Abyss of Time

ARCHAEOLOGY IN SOCIETY SERIES

SERIES EDITORS
Ian Hodder, Stanford University, ihodder@stanford.edu
Robert W. Preucel, Brown University, Robert_Preucel@brown.edu
Rowman & Littlefield Senior Editor: Leanne Silverman, lsilverman@rowman.com

In recent decades, archaeology has expanded beyond a narrow focus on economics and environmental adaptation to address issues of ideology, power, and meaning. These trends, sometimes termed "postprocessual," deal with both the interpretation of the past and the complex and politically charged interrelationships of past and present. Today, archaeology is responding to and incorporating aspects of the debates on identity, meaning, and politics currently being explored in varying fields: social anthropology, sociology, geography, history, linguistics, and psychology. There is a growing realization that ancient studies and material culture can be aligned within the contemporary construction of identities under the rubrics of nationalism, ethnoscapes, and globalization. This international series will help connect the contemporary practice of archaeology with these trends in research and, in the process, demonstrate the relevance of archaeology to related fields and society in general.

Books in this series:

An Archaeology of Resistance: Materiality and Time in an African Borderland, by Alfredo González-Ruibal (2014)

Archaeology beyond Postmodernity: A Science of the Social, by Andrew M. Martin (2013)

UNESCO, Cultural Heritage, and Outstanding Universal Value: Value-based Analyses of the World Heritage and Intangible Cultural Heritage Conventions, by Sophia Labadi (2013)

The Dark Abyss of Time, by Laurent Oliver (2011)

In Defense of Things, by Bjørnar Olsen (2010)

Dwelling, Identity, and the Maya: Relational Archaeology at Chunchucmil, by Scott Hutson (2009)

The Social Construction of Communities: Agency, Structure, and Identity in the Prehispanic Southwest, edited by Mark D. Varien and James M. Potter (2008)

Archaeology and the Postcolonial Critique, edited by Matthew Liebmann and Uzma Z. Rizvi (2008)

Collaboration in Archaeological Practice: Engaging Descendant Communities, edited by Chip Colwell-Chanthaphonh and T. J. Ferguson (2007)

Appropriated Pasts: Indigenous Peoples and the Colonial Culture of Archaeology, by Ian J. McNiven and Lynette Russell (2005)

Archaeology of Performance: Theatre, Power, and Community, edited by Takeshi Inomata and Lawrence S. Coben (2005)

The Dark Abyss of Time

Archaeology and Memory

Laurent Olivier

Translated by Arthur Greenspan

ALTAMIRA PRESS

A division of
ROWMAN & LITTLEFIELD PUBLISHERS, INC.
Lanham • New York • Toronto • Plymouth, UK

Published by AltaMira Press
A division of Rowman & Littlefield Publishers, Inc.
A wholly owned subsidiary of The Rowman & Littlefield Publishing Group, Inc.
4501 Forbes Boulevard, Suite 200, Lanham, Maryland 20706
http://www.altamirapress.com

Estover Road, Plymouth PL6 7PY, United Kingdom

British Library Cataloguing in Publication Information Available

Library of Congress Cataloging-in-Publication Data

Olivier, Laurent, 1958-
[Sombre abîme du temps. English]
The dark abyss of time : archaeology and memory / Laurent Olivier ; translated by Arthur Greenspan.
p. cm.
Includes bibliographical references and index.
ISBN 978-0-7591-2045-7 (cloth : alk. paper)
ISBN 978-0-7591-2046-4 (pbk. : alk. paper)
1. Archaeology and history. 2. Archaeology—Philosophy. 3. Memory. 4. Time. 5. Material culture. 6. Cultural property. I. Title.
CC77.H5O4313 2012
930.1—dc23 2011030580

∞™ The paper used in this publication meets the minimum requirements of American National Standard for Information Sciences—Permanence of Paper for Printed Library Materials, ANSI/NISO Z39.48-1992.

Printed in the United States of America

To my sons:
Rémi, who seeks, and Martin, who invents

CONTENTS

FOREWORD

A few years ago, I was speaking around a table to a group of first-year graduate students about the "state of the field." Without premeditation, responding to something one of the students said (now forgotten), I pronounced rather cavalierly that we were on the verge of a new romantic era. The students, I think, were a bit taken aback. But I pleasantly surprised myself. Perhaps here was an antidote to years of hypercynicism from those too young to have earned such a posture and an anticipatory resolution, or at least productive new phase, beyond the millennial darkness into which we had plunged in the post-9/11 era. Romanticism, long enough rejected because misunderstood as retrograde nostalgia, is due for a return in the wake of too much destructive rationalism. It will not return in the same way, but there will be elements from the early nineteenth century that will seem relevant again. Romanticism, after all, was embodied as much in the man credited with inventing "modernity," Charles Baudelaire, as it was in the wistful landscape drawings of John Ruskin looking back on the Middle Ages as a more innocent time. We have certainly had more innocent times. And we certainly have a sense of entering a new era, as frightening as sometimes that seems. Full of vampires and specters, it terrified Baudelaire as well.

Walter Benjamin's philosophical reflections, a key inspiration and confirmation of the work you are about to read, have been until recent decades repressed in scholarly memory, rejected before their time as too romantic, or *just* romantic. But like the coal miner's lamps that Laurent Olivier tracks as having their own uneven rhythm of formal change, the strain of powerful antipositivism that Benjamin's work preserved in quiet fashion is beginning to be heard again. Most significantly, our sense of time itself is changing. No longer singular and racing like a car down a freeway, time has multiple loops and strands. Few fields are as well placed and as well equipped to explore the time warps of human life as archaeology, if its own practitioners can open themselves up to the romantic critic. This is the opening that *The Dark Abyss of Time* represents— to escape not only evolutionary determinism, but what Benjamin called historicism.

It is hard to imagine this book being written by an American archaeologist, which is why it is all the more important that American archaeologists read it, as delivered by this translation from AltaMira Press that maintains the brave vulnerability of the French original. The legacy of European archaeology that informs this work spans from Roman archaeology to the heavy weight of the Holocaust still haunting the continent's sense of human destiny, with ghosts recently resurrected by a new wave of excavations of twentieth-century European sites that Olivier productively brings to view. On the one hand, we are confronted with a critique of the culture history from which we have never quite escaped. On the other, we are presented with a startling messianic role for archaeology to place the past where it belongs—in the present. This is not romanticism as heritage. Rather it is recognizing that vestiges and trajectories, with a deep, spiraling history, are the stuff of which the present is made—the past is always emerging in the now—or in the case of excavation, erupting. In poetically laying out the potentials of archaeology today, Olivier connects Darwin, Freud, and Benjamin into a new natural understanding of how memory works. This is archaeology as a form of collective hypnosis—a memory recovered of things we may never have otherwise noticed.

Shannon Lee Dawdy
Department of Anthropology
University of Chicago
August 2011

ACKNOWLEDGMENTS

It is impossible to write a book without the direct or indirect involvement of many people. I am referring to my mentors who encouraged me to write. Sander van der Leeuw (Department of Anthropology, University of Arizona) convinced me to take on this project which, initially, was to focus on the history of ideas in archaeology. I am also greatly indebted to scholars whose work has inspired me: Geoff Bailey (Department of Archaeology, University of Newcastle, Great Britain), Richard Bradley (Department of Archaeology, University of Reading, Great Britain), Gérard Chouquer (CNRS, Paris), and Mike Shanks (Department of Classics, Stanford University). Through their knowledge and the questions they raised, I benefitted from them all, each in a particular way. The generous support of my friends, whose interests are only seemingly foreign to the themes broached in this work, has been invaluable, in particular the support I received from Jean-Pierre Legendre (Service de l'archéologie de Lorraine, Metz) and Bruno Wirtz (Department of Mathematics, Université de Bretagne occidentale, Brest). Last, I would especially like to thank Jérôme Prieur, without whom this book would not have been possible.

INTRODUCTION

> In our darkness, there is not
> one place for Beauty. The
> whole of it is for Beauty.
> —René Char, *Feuillets d'Hypnos*

Fragment 237

It is a lovely day. Sunlight fills the air and a refreshing breeze is rippling the grass. I am on a dig with other young people. Walls have been sliced several yards deep into moist sediment with countless red and gray layers filled with charcoal and terra-cotta fragments. We are digging a test trench at the *briquetage de la Seille* in Lorraine, where I worked on my first dig back when I was barely fourteen. The layers of sediment teem with shards clotted with peat, and pieces of bone that gleam a slightly metallic blue. We have hit upon an amazing tangle of stakes, with unbroken vases, wooden objects, and remnants of woven baskets. I am upset because we are driving our spades into this incredibly rich find that looks in certain spots like wrinkled skin, and all we can do is just toss it onto the pile. Now we're sloshing about in silt up to our calves. Objects are falling apart at the slightest touch. The more I try to save them, the worse it gets. We are trampling pieces of wood that crumble beneath our feet, while miniscule pellets with tiny amber pearls are pouring through the openings in our sifters, followed by charred hazelnut shells. Then all the rest of it gushes through in a gigantic black wave that carries us under.

Suddenly I am in a large house that hasn't been cleaned in ages. Dusk is settling very quickly, and all the people working on the dig want to go home, leaving me alone to clean up this huge place with all our equipment and our finds. I am up in the attic, rummaging through piles of household goods and tattered clothes that are supposed to be ours, looking for something that I can't find because I can't remember what I am after. On the other side of a door there is a very dark, small room where we've piled up crates holding what we dug up. The crates are covered with dust and plaster, as if they had been abandoned there decades ago. Their faded labels are completely illegible. In once transparent plastic bags, there is nothing left but unrecognizable clumps of dried-out earth. The owner who lent us this place lives next door, in a house with the lights on. He is a man of considerable stature, and rather distant, because he is dead. Everyone who lives in this place is dead, that's obvious. I don't actually know the location of this property bounded by tall, bare trees, but I can tell that the road that leads up here starts back at a place I won't be able to return to.

This dream expresses the way I experience archaeology far better than I could describe it otherwise. It is the dream version—which is to say, for me, an authentic version—of pre-historian André Leroi-Gourhan's teachings. Leroi-Gourhan in fact says very much just this: that the memory of the past cannot be unsealed without being destroyed by the very act of unsealing it. Archaeologists are people who rummage and demolish. *All that can be had from exhuming some memory of the past is a glimpse of it that is impossible to hold onto, and that dissipates irretrievably.* Everything brought back to us from the past is condemned to fracture and immediately fall apart, because in wresting these artifacts from the memory in which they lie buried, we thrust them back into life, which is to say, to face the lethal assaults of time. And yet, there is nothing else we can do but attempt to draw these vestiges of the past over to the side of the living, to this other side where they will turn to dust, as we will.

Although more often than not they deny it, all archaeologists have at some point dreamed of touching the past up close, of locating that site where a chamber of the past remains intact, just as it was when time sealed it off. If archaeologists were to happen upon some little Earth god who could make their greatest wish come true, they would ask him, as we all would, to grant them, too, a Pompeii or a tomb of Tutankhamun. I would ask for that as well, even though I know that what we are looking for is unattainable.

We are not after lost riches or faraway treasures, as we manifestly lead people to believe. In truth, like Hanold under Italian skies, we are looking for our Gradiva.[1] We know well that she is dead, just as we know that the past is dead and will not return, but we are constantly on the lookout for signs of its ghost. We seek them out under every clump of earth that we turn over, hoping to find traces of it: its bare footprint in the ashes of time, the mark it left when each of its steps imprinted the soil with its presence, its bearing and its gait. And that's all that really matters to us, reaching that ultimate origin of the past when a living presence was making its mark in the earth even as it was fading away, soon to vanish forever. There lives within us all an unspoken dream, a kind of secret belief that something can exist that will not disappear, an imprint that will endure.

The subject of archaeology is nothing other than this imprint of the past inscribed in matter. Fundamentally, it is an investigation into *archives of memory, which is what remains are*. But, what do they point back to? What are the original contexts in which they accumulated over time? These are the principal questions that I have attempted to answer here. The eight chapters of this book are necessarily disconnected and disparate, not unlike archaeological fragments themselves, which rarely fit together. The subject of each of them is as follows:

Chapter 1. By virtue of its subject, the field of archaeology is wholly directed toward origins. Dealing with the origins of archaeological traces inevitably brings us to focus on our childhood, when the attraction to archaeology first took hold, back before the sense that something had been lost evolved into a hunger. The yearning created by the loss of the past brings to light *the paradoxical workings of archaeological time*, which are not linear, as they are in ordinary experience, but pluritemporal, as they are in memory, for they involve several overlapping times.

Chapter 2. We live among vestiges of the past and leave behind what are effectively potential vestiges of our own era. Given that *archaeological finds and the people who find them are inextricably entwined*, how can archaeology generate a field of knowledge specific to it? What vision of the past can this "once upon a time" discipline give us? Having always been overshadowed, in fact thoroughly dominated by history, archaeology has struggled to find *a path of its own, other than that of recounting the past.*

Chapter 3. The basic activity that brings the buried past to the surface is excavation. Here again, the act is paradoxical, for at the same time that digs uncover *vestiges of a past* that had been thought lost, they unavoidably produce *objects that belong to the present.* Once exhumed, artifacts, some of which may come to us from a very distant past, are with us in the here and now, where we attempt to decipher them and reconstruct the history of which they were a part. Thus, what the archaeologist brings to the surface is not so much an object recalling some clearly delimited time period of the past, as the evolving memory of that past, whose meaning can be determined only by and in the present.

Chapter 4. Claiming that artifacts are fully inscribed within the present will surely prompt an obvious objection: Doesn't archaeology, as a "science of the past," deal, almost by definition, with completed periods of the history of mankind? In other words, doesn't there exist a point short of which the objects humans leave behind lack archaeological value, when they are simply old things, things of so recent a past that they don't even belong to history? Recently discovered sites and remains of the 20th century's two World Wars have proven that this is not at all the case. It has become necessary to accredit *an archaeology of contemporary times*, and this proximity has thoroughly unsettled archaeologists' relation to their field. Archaeology effectively deals with every period of time that is recorded within the objects around us, even the most recent of them.

Chapter 5. Artifacts are the focal point of archaeology, and situating them in the present has shaken the discipline to its foundations. Because it is vulnerable to changes in the present that alter our perception of material reality, archaeology has been *affected by the mutations of the contemporary era,* both in its relation to artifacts and in its depiction of the past. The loss of the way we used to experience things, that Walter Benjamin noted in the aftermath of World War I, and that was the result of the industrial cataclysms of the 20th century, has effectively transformed the conditions for gaining objective knowledge of the past. Archaeology, which trades in knowledge of the past, has not been immune to these upheavals.

Chapter 6. The crisis in interpreting the archives of the past that resulted from the unprecedented transformations of contemporary times has effectively stripped artifacts of their standing. Here, in the present, they appear to be residue, or in some deteriorated state—in other words, trash. Thus it is that an archaeologist's work is essentially that of a "ragman of the past," picking up the refuse of times gone by. But a glance at the history of archaeological methodology shows that *the focus has traditionally been on uncovering objects, with little attempt to understand their context.* For a very long time, archaeologists were essentially blind to the process by which traces of the past had been left behind.

Chapter 7. Archaeological sites and vestiges would thus appear to be shaped by a particular kind of memory that lies beyond the grasp of common understanding. In effect, archaeological remains are the product of certain processes of imprinting. Archaeological matter can be defined as *memory-objects that function through reiteration and repetition.* This is the very same process discerned by Freud and Darwin, who used references to archaeological remains in providing keys to the understanding of how the past is imprinted.

Chapter 8. Considering artifacts, as we need to do, as symptoms of a constantly reconstructed memory rather than as objects that bear witness to some past, has opened uncharted paths for archaeologists. Central to our position is the concept of persisting traces insightfully articulated by Aby Warburg. It is a concept that puts archaeology in harmony with its subject, which is to say the material archives of the memory of the past, by *redirecting the discipline's focus to the study of filiations.* Archaeology deals with things as they come down to us transformed over time.

I am well aware of how unsettling such an approach is, not least of all for those of us who are archaeologists. We have waited so long to find Gradiva that we have conjured up our own image of her. By putting together bits and pieces of what we would like her to be, we have imagined her as we hope to see her. And we have seen her. We have seen her in our dreams. We know the stone and the earth she is made of, and the folds in the cloth draped over her. We can just about make out the ancient smell of wood and smoke, the echo of the far-off call of animals, and the clear tinkle of metal. And through the vague traces she has left behind, we see her standing there, radiant in the noonday sun.

In acknowledging that artifacts are evidence of a memory that continues to be reshaped in the present, we abandon our hopes for a past that may some day be seized as it was and find ourselves sucked down into that dark chasm into which ancient times are cast. We run the risk of disappearing into the "dark abyss of time" that Buffon, well before Lyell and Darwin, sensed was opening beneath the feet of the 18th-century man he was. The past has no resemblance to the way we picture it. It rather has the strange appearance of those Gaulish bronze dragons that, until not that long ago, lay buried beneath what are now runways at Charles de Gaulle airport. With their lips curled up under their impressive snouts, they look out at us through half-open eyes, but don't see us. Words emerge from their scowling maws, but we can't understand them, because we can't hear them. But they *are* calling to us.

Note

1. The archaeologist Hanold is the main character in a novel by Wilhelm Jensen that so impressed Freud that he hung the cast of a bas-relief depicting the young woman Gradiva in his office in Vienna. It was some trace of Gravida that Hanold "literally," Jensen says—*im wörtlichen Sinne*—hoped to discover in Italy, and that he eventually found in the features of a young girl in Pompeii. For insight into both the Freudian myth and the concept of *archive*, see Jacques Derrida's illuminating *Mal d'archive. Une impression freudienne* (Paris: Galilée, 1995).

CHAPTER I

IN THE BEGINNING

Childhood

"Long ago, our country, France, was almost entirely covered with huge forests. There were very few towns and, children, the smallest farm in your village would have seemed a palace. France was called *la Gaule* back then, and the primitive people who lived there were called the *Gaulois*. . . . The *Gauloises*, our mothers in past times, were their equal in courage, and followed their husbands into battle. Carts brought along the children and the baggage, enormous, ferocious dogs ran alongside the wagons . . . . "[1]

These words are engraved upon my memory, as if the world they painted had been part of my own childhood. So it is that my attachment to archaeology dates back very far. As a child, I loved stories with castles and knights and wizards. In my mind, somewhere in that vast forest they were lost in, I was right beside those children who came upon man's very first hut in the midst of that clearing they entered before anyone else. I have no idea when those *Gaulois* emerged from the forest of my imagination and began appearing to me in everyday life: on the road that led back home, in the fields and, of course, in the woods.

"If I have placed my autobiography at the beginning of this work," Heinrich Schliemann wrote, "it is not out of some vain sense of pride, but rather to show that the work of my adult life was the natural consequence of things that had struck me in my early childhood, and that the pick and the spade on the digs at Troy and Mycenae were forged, as it were, in the little German village where I grew up."[2]

Throughout his life, Schliemann remained true to the dream he had had as a boy of seven: find the Troy of the *Iliad*, which he was going to search for with Minna, the little girl next door, whom he had promised to wed. But when Minna grew up, she married a rich farmer, having forgotten her childhood vows. Heinrich, on the other hand, never gave up the dream of discovering Troy. "Happy, thrice happy, the man whom fate allowed to fulfill in his adult life a childhood dream, by finding the sacked city of Troy," the German anthropologist Rudolf Virchow wrote in his preface to Schliemann's *Ilios*.[3] Freud, who was awed by the discovery of Troy, went even further. "The man was happy when he found Priam's treasure, because happiness comes only with the fulfillment of a childhood wish."[4]

For many archaeologists, the penchant for probing the past is effectively the product of a childhood desire. I too must have been about seven when my parents first took me to the National Museum of Archaeology in Saint-Germain-en-Laye, where I saw countless broken objects crammed together at the back of tall, glass display cases. Masses of them covered every bit of space available on the cloth-covered panels. Not a single piece was really beautiful, or even attractive. Their bodies were mutilated or deformed, their complexion faded and dusty, and their thick brownish skin covered with enormous scabs and flaked or deeply cracked. And yet, by their simple presence, these discards revealed something unbelievably new to me, and that has amazed me ever since: Something tangible had survived from worlds that had disappeared, something that had belonged to those worlds had come down to us. On that day, those remains became extraordinary for me precisely because they were ordinary. The deterioration they had undergone was the memory itself of the long journey through the darkness of time that they had gone on to get here, where each of us can see them and touch them.

Years went by before I began to find some of my own and learned how to identify them. The very first was a shard that the ice-cold waters of a little forest stream was bubbling over and that I spotted one spring day while my playmates, far ahead, were busy playing with sticks. I picked it up from among the pebbles and held it in my hand. It was a fragment from the bottom of a crock, made of light-colored earth, with a smooth, delicate surface and cracks that had been eroded and rounded by the current. As I slipped it into my coat pocket and carried it home, I knew that I was becoming an archaeologist. An archaeologist: a person who finds things, who resurrects objects from worlds that have disappeared and brings them back to the present, who goes forth with his eyes to the ground where the memory of eras gone by lies buried, who scans the surface of the earth, where time is recorded, in search of traces of the subtle workings of memory.

The Imprint

I have just been told that she passed away at two o'clock. The door to her room stands wide open, and the pale February sunlight is streaming in through the windows that look out onto the garden. She is gone. Her glasses are resting on an open book next to her bed. There is a magazine that was tossed onto a table and a sweater thrown over the back of a chair. To the left, in the closet, neatly put away, are the shoes and clothes she wore when she checked in. Her handbag is sitting on the top shelf. In her wallet are old photos with frayed corners, letters folded in eighths, and some old newspaper clippings. It's a collection of intimate pieces of her life, a smaller, remote version of her home, now occupying one of the few free spaces in her hospital room, as if awaiting her return, which will not happen. All that remains of her is the unmade bed, the imprint she made on the rumpled sheets that hold the memory of her back and her movements from earlier that morning, a fleeting negative that reflects her passing away. A nurse has already come in to tell me that they haven't had time to do the room but that they will get her belongings to me. I won't be able to take most of them. Whatever things I could carry off will turn into dreary, worn remains, shriveled and somewhat repugnant mementos.

The archaeology I hold to insists on dealing with objects themselves. It wants nothing to do with erudite references and elucidations. It is beyond all history, and rejects all scholarship. I want to know what remains of us. I want to know what is in "the true picture of the past," to use Walter Benjamin's phrase. I now understand what he meant when he said that "the true image of the past flits by. The past can be seized only as an image that flashes up at the moment of its recognizability, and is never seen again . . . . For it is an irretrievable image of the past which threatens to disappear in any present that does not recognize itself as intended in that image."[5] The glimpses of the past—which are imprints and vestiges—are indeed intended for us. For them to reveal themselves to us, we have to be prepared to recognize them. I now want to know what appears to us "in a flash" and does not belong to the realm of history.

The archaeology that drives me undermines the conventional notion of historical time. All of time is here and now, not "back then" and "long ago." What remains of the past are the ruins and detritus that time, which is to say the evolving present, is constantly accumulating and breaking down. I want to know what endures in these ruins, what remains inscribed upon the debris once the life that gave it meaning has forever vanished. I would like to know what will remain of us when we perish.

The Black Box

My mother's black lacquered wooden box is now home with me. There is not much to say about it other than that it's an old box with gold-colored Japanese-looking flowers painted on it, that have faded over time. On the top, there is an owl perched on a rock with a little gnarled tree growing out of it. The tree has fruit on it that is still very bright coral red. The minuscule key to the little lock was lost long ago, and one of the hinges has been repaired with an unfolded paper clip.

The box holds objects that have manifestly no relation to each other. Lying on the bottom are rosaries of glass, pearl or boxwood beads, a few glass or metal beaded necklaces, and two gold watch chains, while on top there are two eyeglass cases, one with a pince-nez, the other with a man's pair of horn-rimmed glasses with round lenses. There are also four dried-out, somewhat crumbled cigars, a plaster figurine of a communicant with a pale pink, raffia ribbon wrapped around its legs, keys, and a steel nail clipper in a case made of leather that is all shiny from wear. In with the rosaries and the necklaces there is a citrus yellow plaster pendant with a face that has a flat nose, full lips, and slanting eyes. There is also a plaster heart-shaped medallion, it, too, the work of a child, painted emerald green with a line of unevenly spaced white dots all around the edge.

My memory of this box runs deep. As my fingers glide over the familiar lacquer, my nostrils fill with the distant smell of the varnish. I made it in kindergarten in Versailles, for Mother's Day. I distinctly remember doing the dots with a match tip I dipped into the paint. Beneath the green coloring on the back you can still make out my name, written there back in the early 1960s in my teacher's careful hand. The yellow, plaster pendant was painted by my younger brother, Vincent, also for a Mother's Day, at about the same time. (I remember—or someone told me—that he had it in his mouth when he came home from school, and that he wouldn't take it out because he thought it was a piece of candy. In fact, before I lifted the cover, I thought it was black, like the licorice we used to buy at the bakery back then.)

Along with the lacquered box, I also inherited a tin Muratti's cigarette case filled with other outdated items, most of which are unfamiliar to me: an old woman's watch with a dried-out leather bracelet with cracks, little religious medallions, one of them aluminum with the Infant of Prague on it (it's written right there), a single earring and a pair of cufflinks, yellowing baby teeth, a very old and very worn gold crucifix with fleur de lys at the tips, and another Moroccan (I think) reddish gold one, a large ten-franc silver coin minted in 1967, an aluminum 50 centimes piece from 1942, a broad, yellow-gold wedding band, and a completely twisted little ring with a ruby set in it.

There is a similar assortment of objects in a third little box of gilded wood that is a lot newer. It includes a little black-and-white photograph with scalloped edges of three boys with their arms around each other's shoulders standing in front of a garden wall in the country. There are also three gold or gold-plated brooches, one of them Victorian, set in cotton inside a little black plastic case.

These objects represent the family reliquary that for the most part was put together by my mother and, before her, probably by her paternal grandmother. They represent at least five generations of the history of part of my family on my mother's side. Most of these relics took flight down the road of the 1940 exodus. On foot or by truck, wrapped in blankets, they went from the outskirts of Paris to western France, and were watched over in barns along the way. Then they came back up at the end of the war, after which they were moved from one house to another over the course of a dozen or so relocations between northern and eastern France. These boxes are now in my house. In truth, not much can be said about them other, perhaps, than what Georges Pérec wrote in *Ellis Island*: "You can only try to name things, one at a time, unemphatically, listing them, counting them in the plainest possible way, in the most precise possible way, doing your best to leave nothing out."[6]

Vestiges

Strictly speaking, these objects are vestiges. They bear witness to a history that has vanished, be it real, imagined, or pieced together, which doesn't really matter much. What is more important, I believe, is the basic relation between these objects and the bodies of the people who are now dead. The rings were fitted on fingers, the eyeglasses made marks on noses and ears, the necklaces, crucifixes, and medallions gently lay upon the skin of a neck or a chest. The people who wore these have disappeared, and these pitiful remains, which are pitiful because they are trivial and now obsolete, have become the lone testimony to them, or more precisely stated, their material memory. They are relics, in the same way that artifacts are the debris of what were once very personal objects whose memory has been lost.

The oldest of them is surely the yellow gold crucifix in the cigarette box, and which seems to me to date back to the 18th or maybe even the 17th century. According to my mother, it was handed down from mother to daughter or daughter-in-law. And it was my mother, it seems to me, who ended this practice when she consigned it to this reliquary. The last person to have had it on was undoubtedly her grandmother Marie, who wore it throughout the first half of the 20th century. The rosaries belonged to my great-grandmother Marie's mother, or maybe even, one further generation back, to Marie's grandmother. I don't know the names of these people, but my mother did. These are objects from another time, back when the Catholic religion dominated the lives of women and the family. The glasses belonged to my great-grandfather Paul, as the cuff links probably did. He was given the cigars for his only son's communion, which the figurine of the communicant, probably from the 1920s, commemorates. The pince-nez belonged to his wife, my great-grandmother Marie, as did most of the jewelry, especially the gold brooches. My mother loved to wear them. The snapshot in the gilded wood box is probably of her younger brother Jean-Paul with his friends in a lane near our grandparents' house in Marolles, southeast of Paris. I don't know whose the religious medallions were, maybe my mother's when she was young, or her brother's or cousins', but the beaded glass necklaces and the costume jewelry definitely belonged to her. I especially remember the aquamarine, glass-beaded necklace that she wore when I was a little boy. The nail clipper is my father's. He always carried it in his pocket. The baby teeth are no doubt mine, or my brother Vincent's.

It was primarily my mother, over the course of her life, who put together the collection contained in these boxes from the objects she had inherited and wanted to save or pass on. She occasionally added to the pieces initially given to her by her grandmother, first with our work from kindergarten and probably our baby teeth, then with other, more recent objects she collected over a period of at least thirty years, such as the costume brooches and necklaces that she put in alongside my grandmother's jewelry. My father's nail clipper is probably one of the last items to find its way into the reliquary. It was only today that I learned that my mother had put it there after he died. Maybe the large wedding band is his too; I don't know what happened to his.

In fact, a number of the objects there are a mystery to me. I know neither whom they belonged to nor why they were kept. But I know that some of them are there in place of others. For example, the large silver ten-franc piece replaced a similar one from the end of the 19th century that was kept in a silver mesh coin purse that belonged to my grandmother. I remember seeing that coin purse when I was a child. It probably got lost during one of the moves.

All of the items collected here owe their safekeeping to some specific reason that goes beyond the narrow idea of a keepsake. Taken together, they reproduce a certain sense of our family history as wrought by several generations of women on my mother's side, and as it was handed down among them. The core of this collection originally included objects belonging to my maternal great-grandparents, Paul Félix and Marie, and others that were handed down by Marie's mother and/or grandmother. Nothing comes from my mother's parents, nor of course from my father's; that's because my mother was raised by her paternal grandparents. When her parents got divorced, the four children were separated. The elder two (my mother and her brother Jean-Paul) were placed in their father's care, the younger two in their mother's. This reliquary is thus rooted in my mother's childhood and the reconstructed memory of the generations of grandparents and great-grandparents that repaired the break caused by her parents' divorce. As is the case for any collection, this one is a creation, the reification of a story that successive generations imagined, corrected, and enhanced.

But, at a deeper level, lurking in the substance of time that subtly crystallizes from the several generations of accumulated family objects, there appears another pattern, one that relates their personal history, as the austere religious accoutrements of the old women dressed in black give way to a child's—the child I was—vain attempts at recollection. Other patterns tell the story of the matter itself, the story of the appearance of industrial materials. It is a very long story whose telling is made possible by the large gaps in it, a story told on a scale far greater than that of our individual lives, and whose meaning develops over an expanse of time that we cannot grasp. As parts of history relentlessly keep falling into oblivion, they work away at the objects left behind like an incessant, erosive force, stripping them of their meaning until there is nothing left but mere debris. As Georges Pérec wrote in *Species of Spaces*: "My spaces are fragile. Time is going to wear them away, to destroy them. Nothing will any longer resemble what it was . . . . I shall look at a few old yellowing photographs with broken edges without recognizing them."[7]

The fall into oblivion is the result of change, a consequence of the fact that collections continue to evolve, transformed by new patterns of meaning that replace the old ones as new objects and newly created materials are included. The more we add the present onto the past, the more the past itself becomes matter for conjecture and hypothesis. Who are these people in the pictures? What is this thing? When does it date from? As the intrinsic history generated by the collection of objects emerges, the harder it becomes to consider time a succession of distinct eras, which is the notion of time conventionally associated with history. With objects stripped of their original meaning, it is no longer possible for me to know exactly what period they belonged to. They have all been tossed together, jumbled and entwined. The sense of time through which we conceive history has been thoroughly undone. No longer is it sequential and linear, as the chronological continuity that serves as the thread of history would have it. Henceforth, for vestiges, the sense of time is floating, pluritemporal. In truth, it is uncertain and probabilistic.

Presence of the Past

There can thus coexist different and even contradictory ways of conceiving an inheritance whose material memory is preserved in the present. For any material assemblage to live on, it has to be constantly reworked. My mother, from whom I inherited this reliquary, did precisely that when she added to the collection inherited from her grandmother with more recent objects that had belonged to my father and mainly to my brother and me. I am now faced with a choice: Either I put these boxes on some shelf where I will eventually forget them, in which case the reliquary will be left intact, or I continue to let it function as an object holding our family's memory, in which case I transform it. I have been keeping my older son Rémi's first baby tooth in an envelope. After much hesitation, I decide to put it in the black box. Other relics are now sure to follow. Actually, there was nothing else I could do. For this reliquary to retain its identity, I have to transform it, deform it, and in the end destroy it. That doesn't matter; in fact, it is going to happen anyway. For this collection of remains is evolving outside of what I do with it, like some archaeological construct slowly taking shape, at a rhythm of its own.

The past does not lie behind us, like some older state of things. It lies ahead of us, with us. Like my mother's reliquary, it is a collection of vestiges constantly undergoing change, and our view of it is constantly being altered, at once different and identical. Its history involves repetition, which is simultaneously continuation and interruption. Material creations—objects, sites, landscapes—die once they cease undergoing change. And then our memory of them dies. As the material continuation of the past is handed down, it is necessarily altered. Any history that posits that the past is different from the present, that it is incomprehensible outside of the specific context of its time, is meaningless. It destroys the past.

The black box is now in my house. The mix of hand-me-downs is true archaeological matter with its own structure and its own trajectory in time. Like Marcel Duchamp's "Dust Breeding" that Man Ray photographed, each of my memory boxes is a world of physical remains in miniature, an experimental model of the memory held in matter.[8] The patterns that take shape beneath the surface familiarity of individual objects challenge our instinctive understanding of history. Yet, these are the patterns peculiar to archaeological matter. Here, as opposed to history, events are almost invisible, because they are fundamentally discontinuous. Dozens of years can go by, even a lifetime, before a new object takes its place alongside those already in the black box. Whenever we open these boxes, the items collected inside seem dead, frozen in time, and yet they continue to evolve, constantly changing across a discontinuous time span that extends beyond the scope of our individual existences. The past calls to the future with the familiar little voice of times that are no more. It is saying to us: Don't leave me.

I live with these boxes as we all live in a city or a countryside whose timeless, seemingly motionless history lies beyond our ken. The existence of matter is foremost an insistence, an obstinacy to endure, to be here, to occupy the present with its inert mass. Like my mother's necklaces that have ended up twisted around the rosaries of boxwood beads that belonged to our 19th-century ancestors, matter from the immediate present spontaneously works its way in and around matter from a more distant past, to add to it, not to replace it. Thus, whereas in our conventional grasp of history the present is separated from the past, in my memory boxes the present is mixed in with the past. And it is indeed the past—in this case the material remains of my family's past—that is physically included in the present, right now, at this very moment.

These are two opposing ways of understanding time and history. And the difference between them is a function of whether we take into account how the memory lodged in matter is defined. In the conventional approach, historical time is a series of events. Events are linked to each other and are explained in terms of their relation to each other. But for that other form of history that is comprised of an accumulation of physical objects, archaeological time is one of repetition. Whether it be adding things to a box of memories, or altering the lay of the land, or (re)modeling one's living space, it's the same thing that we are adding to or landscaping or remodeling. The divorce between conventional history and the historical discipline to come, which would be that of the memory lodged in material remains, closely corresponds to the conceptual opposition between historiography and psychoanalysis noted by Michel de Certeau.[9] In effect, as much for archaeology as for psychoanalysis, the world we are entering is at once familiar and disconcerting. It is a world where, as opposed to that of conventional history, past and present are entwined, where events can occur in place of, or in replacement for each other, and above all where events do not act upon those closest to them but rather on others far removed, more or less by intermingling with them. The operative logic is not that of a chain of events, but rather one of matter, of filiation of form, and the creation of memory.

Rue Pasteur

I didn't have much trouble finding the place. It stood alone on the other side of a highway running through what used to be orchards and gardens. I recognized the building immediately: not very big and not very tall. I recognized, too, the alley that runs behind it and the pine tree near the entrance. The cinderblock wall that separated our "residence" from the abandoned orchard we used to play in, and which had since been turned into a housing development, was still there, perfectly intact. As I moved closer, though, I discovered a detail that I had no recollection of, but which occasionally appears in my dreams about this place.

The dream immediately came back to me, as if shot to the surface. The entrance to our apartment house—our entrance—is shielded off by a kind of glass cage that I am trying to get out of. Behind me, the shut-up apartment we'd lived in is in a dreadful state of neglect. The kitchen tiles are covered with broken glass and plaster that have fallen from the walls and ceiling. In the living room, the curtains hanging behind the closed shutters are in shreds, gray from the dust on them. Old yellowing sheets of papers and tattered clothes are strewn over the filthy floor. Pieces of furniture with their drawers hanging open have caved in, and the faded, peeling wallpaper is covered with big brown splotches. My mother and father are in their ravaged bedroom, looking emaciated in their old clothes, going about some incomprehensible chores. They pay no attention to me. They can neither see me nor hear me. They have been dead for years. I would like to speak to them, but it isn't possible. Then somebody or something comes in. So, I run out the open door into the semi-darkness, race down the stairs and run up against this glass cage in the entry that prevents me from getting out, from getting outside where I am standing at that very moment.

CHAPTER 1

It was a relief to be back in the stream of red and white lights on the highway leading me back home. In a short piece entitled "The Cellar," Walter Benjamin says this about remembering:

> The process that determined the way the house that is our life was built has long been forgotten, but when tolls the hour of the assault and the bombs begin to fall, what strange, haggard antique figures emerge from its foundations . . . . During a night of despair, I dreamed of myself madly renewing ties with the very first friend I had at school, whom I hadn't seen for dozens of years and whom I barely remembered, even at that moment. When I awakened, it all came to me clearly: what my despair had unearthed, as after an explosion, was the corpse of the person that had been walled within, so that I might understand that the person residing there now in no way resembled him.[10]

We cannot return to the past, nor can we imagine it from what remains. There is nothing clearly recognizable of it in the present that surrounds us. It is just an empty carcass, and we alone know that it "once was." What remains of it lies buried, like discarded rags and scraps, in our subconscious, in the depths of dreams, far below the surface. And that is all that remains of it, so that dreams become our sole experience of the past—the form of it that actually survives in real time—for it is in the present that the past exists. Only through this filthy, shriveled debris entombed in the ruins of the past can we maintain contact with our own vanished past. But the earliest meaning of the past has been lost, or rather it exists only in a subterranean, fractured form. And it has irreversibly changed over time, which makes it all the more inaccessible, beyond all communication. These forsaken places, scattered with remnants and debris, whose presence continues to have an impact on our present, are the province of archaeology. Just as we do when we dream, archaeology exhumes fragments of the past deposited in the present. Thus, the man who remembers is similar to the man who excavates, as Walter Benjamin so magnificently put it:

> Language has unmistakably made plain that memory is not an instrument for exploring the past, but rather a medium. It is the medium of that which is experienced, just as the earth is the medium in which ancient cities lie buried. He who seeks to approach his own buried past must conduct himself like a man digging. Above all, he must not be

> afraid to return again and again to the same matter; to scatter it as one scatters earth, to turn it over as one turns over soil. For the "matter itself" is no more than the strata which yield their long-sought secrets only to the most meticulous investigation. That is to say, they yield those images that, severed from all earlier associations, reside as treasures in the sober rooms of our later insights—like torsos in a collector's gallery . . . and the man who merely makes an inventory of his findings, while failing to establish the exact location of where in today's ground the ancient treasures have been stored up, cheats himself of his richest prize. In this sense, for authentic memories, it is far less important that the investigator report on them than that he mark, quite precisely, the site where he gained possession of them . . . in the strictest sense, genuine memory must therefore yield an image of the person who remembers, in the same way a good archaeological report not only informs us about the strata from which its findings originate, but also gives an account of the strata which first had to be broken through.[11]

Searching, excavating: that is where it all happens. A man who digs plunges into a buried past. He drives through it and demolishes it with his spade. He is far different from the specialist who likes to collect things like rare coins or ancient tales, and who dreams of maintaining them in their ancient glory, which he believes he alone is privy to. A man who excavates digs down through layers of earth and pages. He pierces and slices and, in the end, extracts only scraps that he doesn't know how to decipher. He turns them over, puts them next to other fragments that they never had any relation to, and gives them meaning they never had. He is well aware that he doesn't know anything about all these things but, conscious of his ignorance, he allows chance to lead him to discoveries. Which is why he is always on the lookout, for he has just so much time. He is indifferent to what no one is supposed to remain indifferent to. He has no respect for what was once revered. Rather, he collects things that have been discarded, that have fallen by the side, or had to be left behind.

Notes

1. Augustine Fouillée (pseudonym G. Bruno), *Le Tour de la France par deux enfants. Devoir et patrie* (Paris: Eugène Belin, 1877), 133–7.

CHAPTER 1

2. Heinrich Schliemann, *Ilios, Stadt und Land der Trojaner. Forschungen und Entdeckungen in der Troas und besonders auf der Baustelle Troja* (Leipzig: F.A. Brockhaus, 1881), 1.

3. Schliemann, *Ilios*, XII.

4. Sigmund Freud, Letter to Fliess, May 28, 1899, in *The Complete Letters of Sigmund Freud to Wilhelm Fliess, 1877–1904*, trans. Jeffrey Moussaieff Masson (Cambridge, MA: The Belknap Press of Harvard University Press, 1985).

5. Walter Benjamin, *Selected Writings,* "On The Concept of History," trans. Edmund Jephcott and others (Cambridge, MA: Harvard University Press, 2003), Vol. 4, 390–1.

6. Georges Pérec, *Ellis Island,* trans. Harry Matthews and Jessica Blatt (New York: The New Press, 1995), 43.

7. Georges Pérec, *Species of Spaces and Other Pieces,* trans. John Sturrock (London: Penguin Books, 1997), 91.

8. Georges Didi-Huberman, *Génie du non-lieu. Air, poussière, empreinte, hantise* (Paris: Éditions de Minuit, 2001), 53–71.

9. Michel de Certeau, *Histoire et psychanalyse. Entre science et fiction* (Paris: Gallimard, 1987), 87.

10. Walter Benjamin, *Sens unique. Une enfance berlinoise. Paysages urbains* (Paris: Maurice Nadeau, 1994), 141.

11. Benjamin, "Excavation and Memory," in *Selected Writings,* trans. Rodney Livingstone and others, Vol. 2, 576.

CHAPTER 2

WHEN ONCE THERE WAS A ONCE UPON A TIME

Once Upon a Time

Ever since man realized that his individual existence was part of a far greater history, he has been searching for the remains of things made by those who preceded him. In the city of Rome, in the very heart of an Empire whose borders were taken for those of the entire human world, the Romans solemnly preserved a small cob and wood hut in the midst of imposing "contemporary" buildings of marble and brick. The poor little relic, it was claimed, dated back to the legendary times of Romulus and Remus, when the city was founded in a wolf den. And the Romans weren't alone. The Greeks of the Roman period held onto vestiges of their Hellenic grandeur with a similar respect tinged with nostalgia. Before them, the Greeks of the Classical Age erected shrines on sites thought to have witnessed Homeric battles that had taken place more than half a millennium earlier.[1]

In preserving the past, in our efforts to inscribe it in the present, we attempt to inscribe ourselves durably in the history of mankind and the universe, well beyond that of our own lifetimes. This explains why, almost from the start, the initial approach to the search for roots went hand in hand with the sense of unshakable power conveyed by monument building. Erecting sepulchers, palaces, and temples was not done simply to create permanent structures capable of withstanding the forces of deterioration; it was just as much an attempt to link them, through the very site on which they were built, to the most remote origins of which they were supposedly the continuation.

This particular relation to the past is not the privilege of classical civilizations alone. As observed by the British prehistorian Richard Bradley, it lay at the heart of monument building in Neolithic Europe, at least as far back as the fourth millennium BCE.[2] We are better informed, obviously, when written texts tell us why kings, pharaohs, and emperors acted as they did. In Mesopotamia, for example, the construction of Babylonian temples is assimilated, almost by definition, to the restoration, perhaps even the re-establishment of what were originally holy sites. For the Babylonians, the past was what lay before us, what was coming toward us. Thus, building and creating something new was essentially equivalent to reactivating what was old and thrusting it back into the immediate present.

One finds an unusual form of this approach to time in ancient Japan, in a rite called *shikinen sengû*. Every twenty years, the imperial shrine in Ise was to be destroyed and then rebuilt exactly as it was. By keeping a holy edifice constantly new, by keeping what was ancient in the here and now, it was spared from aging and deterioration. Cyclical reconstruction arrested the movement of time and turned what was old into an "eternal recurrence." More significantly, it turned the notions of "original" and "ended" into a spiritual concept hovering outside of time, an abstraction reincarnated with every newly built structure. With builders returning to inherited, age-old techniques for assembling and shaping, each generation gave birth anew to what had first existed long before. The spirit of the past was conveyed through the conversion of raw materials, and the shape they took stood there like a sign of its presence.

It is easy enough to see that objects from ancient times are shrouded in mystery. Their original meaning escapes us, for we have no experience of them as they once were. How, then, can we recognize them for what they were after they have been forgotten, short of some cyclical "modernizing" process that can bring them back periodically to the present? Far away from Europe, beginning with the 9th-century Song dynasty, and especially in the 11th and 12th centuries, the Chinese developed a veritable science of ancient objects. Experts knew how to identify ritual vases by their shape and how to decipher the inscriptions they bore. They were able to determine the period and the circumstances under which antiques from the first and second millennia BCE had been made. Catalogues allowed them to identify pieces that the imperial government submitted to them for expertise, thereby allowing Chinese emperors to establish material links between themselves and ancient dynasties that had become legendary.

The Romans, too, went in search of ancient remains. They had archaeologists, such as Varro, whom Cicero referred to as an *investigator antiquatis*, a searcher of antiques. Not a single one of the forty-one *Antiquitates* that Varro wrote in the 1st century BCE has come down to us. We know only that the vast work of erudition of this advisor to Pompeii was intended to provide the Romans living in the final stages of the Republic with full knowledge of their already forgotten past. Cicero said this about him: "We were wandering and straying about like visitors in our own city, and your books led us, so to speak, right home and enabled us at last to realize who and where we were."[3] Roman archaeology was more a study of the past based on ancient writings—what we would call archives—than an attempt to apprehend the past through material remains, which is what we understand modern archaeology to be. It nonetheless involved the (re)-construction of collective identities. By establishing a framework for the way men lived in the past, it informed present-day men of their roots and, in so doing, allowed them to make them their own. As Cicero observed, these visualized reunions with earliest times dealt with family and household life, and brought into play links of filiation and genealogy.

What Is Archaeology?

In Western tradition, the idea of archaeology stems from the Greek notion of *archalogia*, "discourse on old things." For the Greeks, archaeology was above all defined as knowledge of the past, that is, as a record of where they came from. It aimed at reconstructing the history of cities, and the customs and mores of ancient times. This new form of history, which was developed during the 5th century BCE, is to be distinguished from history understood as investigation (or *Historie*), such as that, in particular, found in Herodotus, and of which archaeology is an emancipated form. Whereas history gathers first-hand accounts that are often contradictory and compares them, archaeology brings and fits together *matter* that has often been scattered far and wide. The mission of history, as originally conceived, is to state objectively what happened in the past, whereas archaeology looks to see what the past *evolved from*. In other words, history deals with events, while archaeology is interested in memory. Even though they share a common subject—namely, the past—the two perspectives differ widely. In effect, they diverge not so much by their objectives as by what they base themselves on. As opposed to history, which is a self-contained discipline, archaeology exists only through the material evidence—artifacts—that serves as the foundation of its knowledge. The Romans had it right in speaking of antiquities (*antiquitates*) rather than of archaeology. The starting point for archaeology has always been collecting and collections.[4]

What particular kind of knowledge, distinct from the knowledge that history brings us, can archaeology pretend to? Plato gives us a very clear idea in *Hippias Major*, in the dialogue between the clever philosopher Socrates and the sophist Hippias, for whom knowledge is a trade, and one that pays quite well at that. In fact, Hippias has just got back from Sparta, where the Lacedaemonians had hired him to share his teachings with them. "What did you talk to them about?" Socrates asks, "Astronomy? Mathematics?" "Not at all," answers Hippias somewhat condescendingly, "Those people barely know how to count. Actually, I spoke to them about archaeology, I mean 'the genealogy of heroes and men, the foundations of cities, in short, about antiquity in general [*archaologia*].'"[5] Genealogy, foundations, antiquity: with these earliest words, archaeology linked the present to man's roots by re-establishing lineage and by laying out a succession of periods dating back to the beginning of mankind.

For the initial periods of our distant origins, archaeologists amassed all the available evidence that allowed them to picture how men once lived, how their societies, their skills, and their cultures developed. "That's quite an undertaking!" Socrates exclaims, "If one were to offer a similar picture of Athens, one would have to cite the entire list of men who have presided over the city all the way back to Solon." "You forget," Hippias proudly replies, "that I can memorize a list of fifty names by looking at it just once." He failed to see that Socrates had put his finger on an essential weakness of archaeology: accumulation. As a discipline that is fundamentally descriptive, archaeology can only amass details from the past, bits of information that are self-contained in so far as they are all that we know about the past, since they are all that we have of the past.

Given these conditions, archaeology can have no perspective on the past and can draw no especial knowledge from it. And that suits Hippias just fine, since for him knowledge is a matter of pure erudition, one that deals in concrete facts. All the while flattering him, Socrates cleverly says: "so I understand that the Lacedaemonians naturally enjoy you as one who knows many things, and they make use of you as children make use of old women, to tell stories agreeably."[6] Thus did Socrates lay bare the social function of archaeology: numbing men's consciences by telling them tales, lulling them to sleep with enjoyable, reassuring stories about themselves. The point is clear: Hippias was selling knowledge he did not possess, or rather, he was offering, via the supposed science of the past that archaeology purports to be, an appearance of knowledge that was nothing more than an accumulation of facts. As becomes clear later on in the dialogue, Hippias takes his opinions for knowledge—or, let us say that the way in which he views the world, in this case its origins, is the product of preconceived ideas. In effect, preconceptions and assumptions underlie the accumulation of archaeology's apparently "objective" findings, which explains why we are attracted to it. It appeals to us because it seems to make knowledge of the past accessible by reducing it to a collection of "concrete facts." At bottom, we are attracted to archaeology because it comforts us in various preconceptions that we have about ourselves and of our neighbors as well, who we feel are vastly different from us even though they are fellow human beings.

So, it had actually all been said long ago, well before the boom in archaeological studies that began with the Modern Age and developed even more impressively in the 19th and 20th centuries. The degree to which the discipline has evolved along the Greco-Roman lines of an archaeology of antiquities is stunning; it almost appears to have been its destiny. The archaeology we know today developed first of all from accumulated objects. Thereafter, discoveries of new artifacts that formerly had not been recognized as such forced us to acknowledge the existence of new periods in the "history of the origins of mankind." Little by little, in the 17th century and above all in the 18th, we began identifying ancient objects of a manifestly non-classical style (that is, neither Roman, Greek, Egyptian, Etruscan, etc.) whose growing numbers made it necessary to establish a pre-Roman archaeology called "Gaulish." The frequency with which new artifacts were being discovered increased in the 19th century, reaching dizzying speeds in the 1850s, when the field of today's archaeology was essentially established.

First to be discovered were vestiges of the human race that dated back "before history" and included stone tools, which forced recognition of a "Stone Age" that significantly predated the pre-Roman period. Very soon thereafter, at the same time that the succession of great pre-Classical civilizations in the Mediterranean was being pieced together, a multitude of discoveries, all the more numerous because of what were by then planned digs and research, led to the recognition of a "barbaric age" or "protohistory" situated between prehistory and Roman civilization. Archaeology expanded in a similar manner in the second half of the 20th century to include yet other chronological periods. The archaeology of medieval and modern times came into being as a result of digs and discoveries that unearthed a mass of new archaeological matter from these periods, which until that point had been the exclusive province of history. Less than ten years ago, a similar process began for the archaeology of contemporary times, and even for the very recent past.

Archaeology works by accumulating material remains, and from as far back as the dialogue between Socrates and Hippias, it has never ceased relating man's origins in the form of "agreeably told tales." At the outset, it marshaled the vast quantity of newly found prehistoric vestiges to construct a history of man's triumphant march toward the heights of civilization, which dated all the way back to the primitive times of the Stone Age. As man expanded his control over nature and began settling, he mastered more and more complex technologies—pottery, metallurgy, steel-making—which brought him to the brink of the civilized life that he would attain in the urban societies of Greece and Rome. Underlying the evolution of these civilizations that gradually lifted themselves out of the primitive world of prehistory was the idea of progress, fueled by man's need to free himself from the hardships of nature. According to this view, with each new invention, man increased his control over the natural world and improved his living conditions. There was no need to spell out the outcome of this marvelous story, which still provides us with a narrative outline for the chronology of the pre- and proto-historic periods.

It was implicit in the advent of the contemporary industrial societies of the 19th century that had at last brought to completion man's attempt to dominate nature, an attempt that had begun in the darkness and vulnerability of prehistoric caves. Like the gods of antiquity, the 19th-century archaeologist seated in his study could contemplate from the pinnacle of his civilization the development of a process from which he saw himself detached. The evolution of mankind coincided with man's growing intelligence, culminating in science.

The history of human progress that was read into prehistoric artifacts was thus intended to comfort the educated 19th-century European in his belief in the absolute superiority of his civilization and its technology. Inventing history from artifacts was carried to an extreme by 20th-century totalitarian regimes that modified the meaning implicit in the conclusion. No longer was it a somewhat unpredictable consequence of progress that led man to dominate nature and his fellow man; it was now a demonstration of the superiority of one particular race, to which the whole of the history of mankind could be reduced. The educated European of the 19th and 20th centuries found himself at its culminating point precisely because he belonged to the "superior race," history having demonstrated that it was his birthright to dominate mankind by virtue of his exceptional capabilities.

Archaeology could demonstrate that, from prehistoric times on, all of the most significant innovations had been the work of particular civilizations of a single biological stock, called Indo-European or Indo-Germanic. Archaeology could even determine the original lands on which this superior European race had been developing since the dawn of time and where, unfortunately, the vagaries of history had allowed civilizations of other "inferior races" to gather, like so much trash, and of which this *Lebensraum* had to be cleansed. The history told by archaeology was no longer merely flattering; it had become a weapon for the enslavement and annihilation of peoples so unfortunate as not to belong to that superior "noble race" of Europeans, specifically, the Germans. This evil tale, whose inevitable conclusion was the industrial-scale death camps of the Third Reich, was abruptly interrupted in 1945. As often turns out in fairy tales, the sweet grandmother who was telling us these stories was actually a wicked witch with a penchant for destruction and death. There is now good reason for us to be suspicious of all such tales.

CHAPTER 2

The Recalcitrant Past

Having undergone unprecedented expansion during the hundred years from 1850 to 1950, archaeology now finds itself in a period of crisis that reaches down to its very foundations and which became apparent during the post-war years. The crisis is first of all quantitative, a result of the amount of artifacts that have been amassed. Museums and depositories are so crammed with archaeological finds that certain experts are seriously considering returning them to where they were found once they have been identified. There are too many sites for us to be able to preserve and study them all, and too many areas of specialization for us to be able to master them all.

Today, working alongside archaeologists themselves one finds ceramicologists, anthropologists, archaeozoologists, geoarchaeologists, palynologists, and archaeobotanists, as well as dendrochronologists and geophysicists.[7] The exponential accumulation of archaeological matter has produced a corresponding increase in areas of study, and thereby in the number of specialized examinations of artifacts. The principal problem facing archaeology today is the uncontrollable mushrooming of these fields and the impossible task of gaining comprehensive knowledge that goes beyond the simple juxtaposition of isolated observations. In spite of the late 19th-century attempts of Gabriel de Mortillet, who wanted to federate the new disciplines of Prehistory, and those made in the period between the wars by Henri Hubert, who sought to redefine the archaeology of proto-historical periods, a cross-disciplinary approach goes wanting.

Archaeology has also been directly affected by the crisis in recounting history that was provoked by the totalitarian regimes of the 20th century and that reached across all fields of history in the latter half of the century. History, whose mission it is to re-establish the truth about the past, did not emerge from this period unscathed, for it was clearly demonstrated that it was well within the realm of possibility to invent a wholly fictitious past and to substitute it for the "old" one, which was to disappear.[8] For the past to fall into oblivion, one needed only to methodically eliminate all of the "physical bearers" of its memory and destroy all physical traces of it. Doing so not only made it possible to "cap" the past, but also went one step further by doing away with it entirely and making its very existence appear unimaginable. Archaeology, whose fundamental practice is that of producing "material facts," found itself directly involved in this perversion of history and was contaminated by it.

Nowhere in Europe had archaeology ever known the growth it experienced under totalitarian regimes and especially under the Third Reich.[9] The findings of the 19th century were completely reworked, and a new methodology, which still serves as the foundation for archaeological studies, was developed. In fairy tales, wicked witches never die. They get locked inside some object and wait there for something to break the spell that imprisoned them. That threat still hovers over us.

In the 1960s and 1970s, the crisis in the recounting of history generated a "new social history" that challenged the traditionally event-based discipline, but no such change occurred in archaeology, which from the very outset has always equated knowledge of the past with the identification of artifacts.[10] And yet, the crisis that has affected all disciplines that seek to learn about the past actually runs deeper here than anywhere else, for it dovetails with the crisis in interpreting objects that began at the turn of the 20th century and that has been deepening ever since. Not only do we not know how to fit together the fragmented areas of knowledge generated by the increase in approaches, we are actually no longer clear about how to interpret the deposits we come upon. The reassuring certainty we once had keeps slipping further and further away. Today, we are not at all sure that the remains from ancient civilizations, which used to seem so obvious to us, are actually related to what they are supposed to be evidence of.

An example taken from a literary work by Georges Pérec allows us to grasp, in a round about way, what this is about. In Pérec's novel, *Life: A User's Manual*, he juxtaposes a large number of stories involving the residents of a Parisian apartment house. We find ourselves in Geneviève Foulerot's bedroom, faced with a particularly knotty mystery to which, let's make it clear right off, there is absolutely no solution. The large painting there by her grandfather Louis is entitled *The Gold Fish Murder*. Its source, Pérec tells us, was a detective novel that "gave him such pleasure as to make him think of using it as the subject of a picture which would bring almost all the elements of the mystery together into a single scene."[11] And indeed, all of the clues to the murder of the rich Swiss diamond dealer Oswald Zeitgeber—who was killed by several people while he was attempting to commit suicide!—are depicted. The weapons are right there in the bedroom in which the murder took place: the gold fish bowl, the old school desk, the water pitcher, and, of course, the wall phone. All of these objects are clearly shown, but we are unable to connect them because we don't know the implausible story that they are a part of. Somebody (Pérec, in this case) has to tell us, because there is absolutely no way to figure it out. Even though it is a perfectly logical story, it is impossible to deduce it from the clues, for we know neither what happened nor, above all, the reason things happened as they did.

Ethno-archaeology is the study of objects produced by present-day societies that we deem to be closely related to ancient archaeological civilizations. It demonstrates rather often that when dealing with vestiges of ancient societies we are confronted with an impenetrable mystery of the same order as that of *The Gold Fish Murder*. Work done by the British archaeologist Ian Hodder on the Nuba of Sudan offers an illuminating example.[12] The Nuba population is divided into two groups, the Mesakin and the Moro. Both groups live in enclosed spaces that they inhabit during the same periods of the year, and both farm and raise animals. But, as opposed to the neighboring Moro, the Mesakin Nuba let the inner courtyards of their living spaces fill up with detritus. Where they live, Hodder reports, "the ground is covered with layers of animal droppings, broken crocks and gourds, and the usual range of household garbage."[13] The average Western explorer would conclude that, unlike the Moro, the Mesakin belong to a native population unaware of the health risks represented by decomposing waste exposed to sunlight. A no less average archaeologist excavating these sites centuries after they had been abandoned would surely conclude that there were two distinct activities going on, one in the living spaces inhabited by the Moro, the other in those of the Mesakin. For him, the waste-covered ground would indicate that the inner courtyards of the Mesakin had been a beehive of activity, whereas he would conclude the opposite for the Moro.

The explanation is actually quite different. The surrounding walls of the inner courtyards of the Mesakin are painted with elaborate symbolic patterns intended to purify the activities of daily life. People can blithely throw their garbage into the courtyards because the paintings protect them from contamination. But in Moro culture there is nothing similar relating to refuse; there are no paintings around their living spaces. They keep them clean by sweeping out filth and by keeping domestic animals from roaming around so as to keep their courtyards free of excrement.

And it is not only with regard to refuse that our preconceptions would lead us to misinterpret Mesakin behavior. For example, when the Mesakin eat the meat of domestic animals, they throw the beef bones in one place and the pork bones in another. In fact, the animals are slaughtered, butchered, and prepared in carefully separated spaces. Were we to encounter similarly segregated butchering areas on a dig, one for pigs, one for cattle, our common sense would lead us to conclude that there existed distinct activities that we would surely interpret in a social sense. We would see signs of different groups that raised different animals, or social classes that did not have access to the same foods. Here again, the real explanation for this behavior would escape us. As Ian Hodder explained: "in Mesakin society, the two domestic breeds symbolize the two sexes. Bovines and their milk are associated with the male's strength and virility, and women are not allowed to milk cows. Pigs, on the other hand, are considered filthy and thus associated with women. Generally speaking, Mesakin men consider women to be potential bearers of germs, especially during menstruation, and they eat apart from the men. Separating pork and cattle bones is a symbolic extension of the separation between the sexes."[14]

Most often, the significance of the unearthed objects we come upon escapes us because we know nothing of their real history. Why, then, is archaeology, as a form of history, so despairingly superficial? It is really too bad that Hippias didn't think to put the question to Socrates, but then again, Hippias was not a man to question things. Let's have a try at an explanation. Archaeology never truly succeeds as recounted history because we fundamentally misconceive the nature of the objects with which we work. Because archaeology studies artifacts that have survived from the past, it actually deals less with what ancient times were like than with what happened to these man-made things. This is nuance, to be sure, but one that opens a huge gap. In effect, it is not because objects belong to the same past—or if you prefer, to one particular period of history—that each displays the distinguishing traits that separate that particular past from all others. Quite the contrary: The physical matter of human societies is comprised of a mass of disparate objects that serve different purposes and which, above all, do not all carry forward the same heritage, whether in terms of technique or of style.

We can get a good sense of the nature of the problems involved in interpreting remains through another instance in which it was possible to compare archaeologists' interpretations with what actually had taken place. In the early 1970s, a group of anthropology students from the University of Alberta, led by Robson Bonnichsen, performed an unusual experiment that became well known in American archaeological circles as "the Millie's Camp experiment."[15] The goal of the experiment, which would find failure at every turn, was to measure the degree of accuracy of archaeological reasoning and interpretation, which is impossible in normal circumstances because the people who inhabited dig sites have long since disappeared. The idea was for the researchers to use archaeological methodology to study a campsite that had been vacated only a short time before and to determine the way society functioned on the basis of the deposits found there. Following that, they would submit their conclusions to one of the former occupants who would tell them whether they had accurately analyzed and interpreted what they had found. The site chosen was an Indian campsite in the Canadian Rockies where Millie, a Cree woman whom Bonnichsen had met while camping in the area, had lived shortly before.

Millie's Camp was a veritable gold mine for archaeologists. Millie's party had made fires, slaughtered and butchered animals, covered the ground with a wide variety of detritus—they counted more than 150 kinds—and even carved symbols in the bark of nearby trees, as well as the dates they had been there, which stretched over a period ranging from 1956 to 1969. The research group examined the grounds carefully and then submitted their findings to Millie. She was at first intimidated by their strange questions, but eventually went along with it all cheerfully. While the group was able to recognize the existence of several areas of activity within the campground used by different people, they misinterpreted what they were. They thought that several families had been living there, whereas only Millie and her two youngest children had. They saw signs of daily activity in waste that had been left behind periodically by Millie's husband and two elder boys, who would come home on weekends after their week's work in the mines. Nor did the group see that certain places in the camp had occasionally been used for different purposes, in particular when wild game was being butchered. For example, whenever fires were made for smoking meat, the areas used for preparing food were moved close to them for the sake of convenience.

The most remarkable misinterpretations involved objects. The research group never imagined how widespread the practice was of using tools for purposes for which they were not intended, or of recycling objects. For example, they imagined implausible hunting activities on the basis of strings from sticks of dynamite that Millie's children played with. They thought long and hard about the intended use of what were just objects fashioned out of branches by Millie's youngest, took improvised devices used for stretching deer hide for snow sunglasses (!), and failed to realize that the mysterious "pictograms" carved on the trees around the camp were the work of Millie's daughter, who was entering adolescence.

Chance, too, played tricks on them, and they were unable to see through it. Millie's Indians had no horses, but the researchers thought they did because of manure they found on the site. In fact, it had come from a horse that had passed through either before or after Millie. Similarly, on the spot where Millie's oldest son had set up his tent, they found objects—most notably a hair curler and a pink plastic whistle in the shape of a bird—that never belonged to him. They had been left there, no doubt unintentionally, after she had gone. Still, it was enough to make the research group believe that there had been several women living on the campsite. In other words, they attributed to Millie's group objects that had absolutely no connection to it because it was impossible to recognize them as foreign.

The world of physical objects is in fact far less obvious than we tend to believe. The Millie's Camp experiment demonstrates that, paradoxically, we are at a loss to situate objects in time. For, once they have been made, they live on usually well beyond the point when the moment in history that saw them come into being has faded into oblivion. What happens to them at that point? Not only do they doggedly remain, like old bodies in the physical world of the presents that follow theirs, above all they end up incorporated into them. That is the lesson of the makeshift and recycled items, and all the things thrown together at Millie's Camp. Note, too, that what happens to objects happens to places as well. The streets of our

European cities very often follow paths dating back to the Middle Ages, or even Roman times. In the United States, highways still follow trails traveled by the first European settlers in the 18th and 19th centuries, who themselves had followed those used by native Americans and perhaps even paths that had been cut across the land by the timeless migrations of buffalo herds.

What we inherit from the past rarely comes down to us as it was. Rather, things are reinterpreted, so to speak, repeatedly used in unexpected ways, in a present they had not been intended for. By the same token, being constantly "thrown back into play" allows things, often very old things, to carry on into the present where they are reworked and enhanced. Thus, far from being explicit, as one might think at first glance, the history they transmit is implicit, discreet, never immediately apparent. Because it deals with objects that come to us from the past, archaeology compels us to conceive of history in a different way: as the memory things hold.

A Discipline in Search of a Theoretical Base

Thus, at the heart of the problematics of archaeology, which essentially deals with memory recorded in artifacts, there lies the matter of time, or more precisely, temporality. But we still don't know how to conceptualize the knowledge peculiar to archaeology. "Archaeology," said the British archaeologist David Clarke, "is an undisciplined empirical discipline. A discipline lacking a scheme of systematic and ordered study based upon declared and clearly defined models and rules of procedure. It further lacks a body of central theory capable of synthesizing the general regularities within its data in such a way that the unique residuals distinguishing each particular case might be quickly isolated and easily accessed . . . . Lacking an explicit theory defining these entities and their relationships and transformations in a viable form, archaeology has remained an intuitive skill—an inexplicit manipulative dexterity learned by rote."[16]

David Clarke stated it well: The lack of clearly formulated theory has kept us from creating a framework for systematically dealing with archaeological remains and the transformations they undergo over time and space. Yet, this is what needs to be done. And there should be no reluctance, since the discipline is still in its infancy, to draw on intuitions and illuminations from the world of art. Writers and artists are often able to depict phenomena that their contemporaries in the sciences can only approach and not yet theorize fully. In this regard, literature and photography have a privileged relation to memory and traces of the past, which we have referred to here as *vestiges.*

Archaeology is still in its very earliest stages as a discipline. As a study of artifacts, it has existed for only a hundred and fifty years, and vast domains—such as the archaeology of recent periods—have yet to be explored. One has only to think of the history of physics, an example *par excellence* of the so-called hard sciences, to see that it took more than two thousand years before that discipline was able to recognize what had to be studied and generate meaning of its own. It was on the basis of repeated experiments with weights dropped from the tops of towers and inclines that men came to define the parameters—mass and speed—with which to build their knowledge of matter and the universe. The whole of Newton's physics rests on these two variables, and all of contemporary physics rests on his. Until then, physicists resembled archaeologists in that they expatiated on nature. They debated the essence of time, the nature of matter, and the order of the universe, just as archaeologists today debate origins, the essence of civilizations, and the nature of the processes of cultural evolution.

Like their pre-16th-century precursors, archaeologists do not yet know enough about the objects they need to study from among the mass of archaeological matter; they still lack meaningful parameters with which to establish a formal structure for archaeological phenomena and, thereafter, an overall theory. Identifying the focus of physics did not really change the subject of discussions about the nature of the universe; it simply allowed physicists to speak about it differently and to observe what had until then been abstractions. They were led to examine phenomena that until then had been matter for discussion or stories. Similarly, directing energies toward the formulation of a theory does not mean that the study of artifacts is to be left behind as we wander off into so-called theoretical considerations. Quite the contrary, the point is to determine the focus that will allow us to build a new kind of knowledge of the past that is more than commentary.

Archaeology's primary task, to borrow the American archaeologist Lewis Binford's phrase, is "decoding the archaeological record."[17] If, because they are immersed in the present, artifacts cannot bear witness directly to the past, of what are they an expression? In other words, if it is not the past, strictly speaking, that lies at the origin of archaeological matter, what does? Archaeological matter is material memory, and memory is a property of everything that comes to life, evolves, and disappears. Like everything that lives and dies, archaeological matter is torn between ephemerality and repetition. Our existence is transient, and our material creations are as well. There is nothing of ourselves, nothing of the objects we use or the spaces we inhabit that is destined to last, or even remain intact, at least not as it originally was. Every moment of time irremediably alters or deforms what exists, and there is nothing we can do about that. Birth, growth, aging, and death are everywhere entwined. In the end, all that will remain of us are pieces of used furniture, piles of old clothes, unmatched dishware, and yellowed sheets of paper. Only scraps will remain, that time, charitably, will eventually break apart, leaving vestiges, traces. The world is already so saturated with the past that the present hardly has a place of its own. Once its time is up, everything here has to die, as quickly as possible, to make room for what is coming. What exists must disappear, and what stubbornly persists must be forgotten: Those are the basic requirements to be met if the people and the things that make up this world collectively are to continue to exist and perpetuate themselves beyond the individual disappearance of its inhabitants.

We live and die in the present. The actual time span of our history does not exceed a lifetime, perhaps just a few decades, which we cannot depict as they were actually lived. The rest is simply what we can decipher or piece together; in a word, what we can imagine. Our link to the past is made of objects, of heterogeneous artifacts that we have had no experience of and which we will never experience as they originally were, if that means anything. In truth, we know nothing at all and will never be able to know anything about the past "in the past," when it was the present unfolding; it has passed by and gone. At best, there may remain some scraps of it. Given this situation, the interpretation of vestiges becomes problematic. The only knowledge of ancient times that we can acquire through them is relational; that is, a knowledge based on our particular relationship, here and now, with the remains of the past that it is given to us to deal with.

And that is a major difficulty of which the traditional approach to the past feigns ignorance, postulating rather that it is possible to reconstruct the series of periods that the past has gone through. But can that be so? The problem posed here involves the nature of artifacts: What kind of information have they recorded? How have they been transformed over time? And what can they enlighten us about today?

Archaeology's Failed Revolutions

Archaeology is a fickle discipline that will never be entirely in our grasp. Like all historical disciplines—all those that deal with the past, either the past of men or of nature or of the two together—it is a consummate "once upon a time" discipline. Events occur only once, and once they have happened they can never be re-examined, as opposed to phenomena studied by the so-called exact sciences for which the possibility of reproducing experiments is the very basis of their knowledge. We have to learn how to go about things differently.

Like other historical disciplines, archaeology deals in traces and fragments. The remains it brings into the light represent but a fraction of what once existed. The vestiges it unearths are all fundamentally incomplete and truncated. They are deformed accounts of a past whose actual historical pertinence remains, in truth, uncertain. Archaeology disinters fossils. The term comes from the Latin verb *fodere*, meaning to dig. Strictly speaking, a fossil is anything that is dug up from the earth.[18] Given these conditions, one might think archaeology doomed to inconsistency, but it has a great strength, one we have little control over: It reveals that which remains from the past and is buried in our present. And, in doing so, it unveils that which is unknown, or has gone unrecorded, but which is lodged in our world, which is a world we would prefer to conceive of as essentially known, or at least entirely knowable. Archaeology shows us that this is not so. The soiled, broken remains that workers and farmers discover as they dig into the ground come from worlds that are foreign to us.

Because it must constantly deal with what is incomplete, uncertain, and unknown, archaeology is fundamentally given to fabrication. It has to fill in huge gaps in the past, whose physical remains are just a fraction of the whole. And it has to reconstruct, somehow, the history that the remains it uncovers were part of. Archaeologists find themselves torn, as has been the case from the outset, between two opposing desires. On the one hand, they are fascinated by the idea of disclosing what once lay undiscovered and unexplored, while on the other, they are irresistibly drawn to make what they unearth a part of what is already known. The wavering between revolution and assimilation has generated tension and brought into play what Bruno Latour would call the discipline's efforts at mediation.[19]

In fact, from its beginnings in the second half of the 18th century as a branch of knowledge involving field work, the history of archaeology has been far from that triumphant march toward objective knowledge of the past that the traditionally positivist historiography of the discipline is so fond of portraying. Its evolution across the centuries has been marked by what can be called the discipline's three "failed revolutions," which date back to the 18th century. On three separate occasions, archaeology almost lifted itself out of the rut in which it was stuck, only to fall back in.

In the 18th century, archaeologists began uncovering—with what were then called "Gaulish antiquities"—the existence of an unknown past of which the classical historic tradition had absolutely no record. The 19th century took things further with the discovery of artifacts from what archaeologists decided to call "prehistory," that is "history prior to history." Very few authors—with the possible exception of Sigmund Freud—recognized the significance for history of this radical change in framework. No longer was history the entire past. It was merely a fragmented representation of what we were aware of, while there lay buried in the earth a vast physical memory bearing witness to what would henceforth be known as "prehistory," that is, a primitive past that had for the most part escaped our historical conscience. These stunning revelations were immediately co-opted, assimilated into historical narrative, whereby "prehistory" became the extension into the past of periods for which there was no written record. In effect, the archaeology of prehistorical times became a subdiscipline of history, with the aim of writing a subhistory of past civilizations that would be compiled, not on the basis of written texts, but more prosaically on that of physical remains.

In the 19th century, archaeologists developed—with what Scandinavian archaeologists called *typology*—an approach to time that was peculiar to archaeology: time as it is recorded in the physical remains of the past. Some writers, such as the Swedish archaeologist Hans Hildebrand, immediately saw the connections with the natural sciences; for example, with the revolutionary framework implicit in Darwin's theory of evolution, which could be used to explain the historical dynamic of mankind's past and present material cultures. At this point, archaeology could have freed itself from the servitude of history, since it now had its own sense of time, one that was not that of the discipline that had made of it its handmaiden—but, once again, these startling discoveries were co-opted. In the hands of an explicitly historicist approach to the events of the past, the typology intended for archaeological matter became a tool with which to construct a linear time for prehistory.

Last, in the 20th century, with the advent of reliable dating methods, such as carbon dating and dendrochronology, archaeology demonstrated that there is a natural time recorded in the physical remains of the past. Very few writers saw, as the British archaeologist Colin Renfrew did, that the introduction of exact dating methods meant certain ruin for the patterns of cultural evolution that had until then been drawn from artifacts of ancient societies.[20] Relocating archaeological remains in their proper place in irreversible time often revealed huge discrepancies with archaeological dating, and gave the lie to the historical accounts into which ancient remains had been incorporated. Of deeper significance still, using scientific dating methods to establish an archaeological time that had, until then, been purely relative opened a breach that revealed the fundamentally probabilistic nature of the information contained in artifacts. Radio carbon dating, in particular, does not function by dates per se, but by estimated periods of time that depend on the amount of carbon-14 remaining in tested objects.

Why have revolutionary discoveries always succumbed to the forces of conformity throughout the history of archaeology? What is it that keeps archaeology a prisoner to archaic schemas of the past and of history? With the crisis in interpretation confronting archaeology today, the need for a new theoretical foundation has become increasingly clear. The sense of time that archaeologists draw from their conventional understanding of history has been shown to be in contradiction with the sense of time that they have found recorded in artifacts. The crux of the matter is the way history is to function, not so much by defining what can be said about the past as by determining what the past in itself can tell us. The German philosopher Walter Benjamin paved the way with his radical critique of the conventional approach to history, in particular in the theses in "On the Concept of History," which he wrote several months before committing suicide in 1940.[21]

Benjamin determined that the past receives its identity in the present.[22] Today, his writings are considered immensely important for their contribution to the philosophy of 20th-century history. In the last few years, their genius has begun spreading through the field of art history.[23] The work of another German, the "art anthropologist" Aby Warburg, who was as unorthodox as Benjamin, is now considered equally essential for the study of contemporary anthropology and art history.[24] His research focused on Greek and Roman traces surviving in Renaissance art works, but the work he did as an art historian was thoroughly undermined by his encounter with the rituals of Indians living in the American Southwest in the 1890s.[25]

We are now at a point when we can embrace the thinking of these intellectuals who defy classification—philosophers? art historians? anthropologists? archaeologists?—and grasp the implications that directly concern archaeology, for their work focuses on the memory of the past as reproduced in material creations. In all events, the traditional concept of history, for which Benjamin used the term "historicity," appears completely outdated. Archaeology does not (re)construct a "history before history"; it deals with the material memory of the past that escapes historical consciousness. Artifacts have no truck with historical time. More precisely still, they resist all attempts to incorporate them into classical history. Once this has been understood, we begin to see the rich connections linking archaeology to other disciplines that deal with memory: on the one hand, the Darwinian theory of evolution, which deals in its way with the "natural" memory of species, on the other, Freudian psychoanalysis, which deals with the memory of the psyche.

The past understood as material memory: that is what archaeology brings to light, and what the traditional perception of history staunchly refuses to accept. Those who hold to the conventional historical approach to archaeology—and they hold to it all the more tightly because their grasp of archaeological matter is weak—know well that this means an "end to the (hi)story." By "the past," one also has to understand the present, since the past is the very matter of the present, its thread and its substance. The present as physical object can be seen only fleetingly, during times of catastrophe, which we always rush to put behind us. These rare passing moments when daily surroundings are abruptly reduced to rubble show us that we too are things: bodies become remains—which we ourselves never see—just as the objects we handle every day without ever seeing them as things suddenly become vestiges. Obviously, "archaeologizing" life allows us to see something that is not normally visible, something we have no name for and which essentially escapes us. That something is archaeological matter. It provides archaeology with a meaning that is peculiar to it as a discipline given to the study of reality, our reality, based on its materiality. And that is the direction I would like to take us in now.

Notes

1. Susan E. Alcock, *Archaeologies of the Greek Past. Landscape, Monuments and Memories* (Cambridge: Cambridge University Press, 2002).

2. Richard Bradley, *The Past in Prehistoric Societies* (London: Routledge, 2002).

3. Cicero, *Academica,* in *Cicero, De Natura Deorum, Academica,* trans. H. Rackham (London: William Heinemann, 1933), I, pp. iii, 419.

4. Krzystzof Pomian, *Collectionneurs, amateurs et curieux. Paris-Venise XVIe-XVIIIe siècle* (Paris: Gallimard, 1987); Alain Schnapp, *La Conquête du passé. Aux origines de l'archéologie* (Paris: Carré, 1993), p. 12.

5. Plato, *Hippias Major*, 2 85e, trans. Benjamin Jowett, http://www.ellopos.net/elpenor/greek-texts/ancient-greece/plato/plato-hippias-major.asp?pg=8.

6. Plato, *Hippias Major*, 2 86a.

7. Ceramicologists study archaeological ceramics; anthropologists and archaeozoologists the archaeological remains of humans and animals. Geoarchaeology (the study of the impact of humans on sedimentary formations), palynology (the study of pollen), and archaeobotany (the study of vegetal remains) are disciplines in the field of paleoenvironmental studies. Dendrochronology is a dating method based on the growth rings on tree trunks, whereas geophysical prospection allows us to locate buried archaeological sites through the physical anomalies they generate in the soil.

8. Hannah Arendt, *Between Past and Future; Eight Exercises in Political Thought* (New York: Viking, 1972).

9. Jean-Pierre Legendre, Laurent Olivier, and Bernadette Schnitzler, *L'Archéologie nazie à l'Ouest du Reich* (Paris: Infolio, 2007).

10. Lucien Febvre, *Combats pour l'histoire* (Paris: Armand Colin, 1953); Fernand Braudel, *Écrits sur l'histoire* (Paris: Flammarion, 1969); Jacques Le Goff, *La Nouvelle Histoire* (Paris: Complexe, 1988).

11. Georges Pérec, *Life, A User's Manual*, trans. David Bellos (Boston: David R. Godine, 1978), p. 222.

12. Ian Hodder, *Symbols in Action* (Cambridge: Cambridge University Press, 1981).

13. Ian Hodder, "Éthnoarchéologie: une approche conceptuelle" in *Les Nouvelles de l'Archéologie*, 1980, vol. 4, p. 28.

14. Hodder, "Éthnoarchéologie," p. 28.

15. Robson Bonnichsen, "Millie's Camp: An Experiment in Archaeology" in *World Archaeology*, 1973, 4, 3, pp. 277–91.

16. David L. Clarke, *Analytical Archaeology* (London: Methuen, 1978), p. xiii.

17. Lewis Binford, *In Pursuit of the Past* (London, New York: Thames & Hudson, 1983).

18. According to the definition that appears in Élie Bertrand's *Dictionnaire des fossiles propres et accidentels* (1763), fossils are "everything either drawn from the earth or that is lodged in it." As a matter of fact, the first drawings of fossils from the 16th century, such as those on the plates in a work by the Swiss naturalist Conrad Gesner dating from 1565, lump together crystals, stone tools (such as perforated Neolithic stone axe-heads), concretions, fossilized fish teeth, sea urchins, etc., on the basis of the amount of geometric detail.

19. Bruno Latour, *Nous n'avons jamais été modernes. Essai d'anthropologie symétrique* (Paris: La Découverte, 1989); *La Science en action. Introduction à la sociologie des sciences* (Paris: Gallimard, 1995).

20. Colin Renfrew, *Before Civilization. The Radiocarbon Revolution and Prehistoric Europe* (London: Cape, 1973).

21. Benjamin, *Selected Writings,* "On The Concept of History," pp. 389–400.

22. As evidenced in his unfinished work, *The Arcades Project* (Cambridge, MA: The Belknap Press of Harvard University Press, 1999).

23. Marie-Cécile Dufour-El Maleh, *La Nuit sauvée. Walter Benjamin et la pensée de l'Histoire* (Bruxelles: Ousia, 1993); Françoise Proust, *L'Histoire à contretemps. Le Temps historique chez Walter Benjamin* (Paris: Éditions du Cerf, 1994); Sylviane Agacinski, *Le Passeur de temps, Modernité et nostalgie* (Paris: Le Seuil, 2000); Georges Didi-Huberman, *Devant le temps. Histoire de l'art et anachronisme*

des images (Paris: Minuit, 2000); Michael Löwy, *Walter Benjamin. Avertissement d'incendie. Une lecture des thèses "Sur le concept d'histoire"* (Paris: Presses universitaires de France, 2001).

24. Georges Didi-Huberman, *L'Image survivante. Histoire de l'art et temps des fantômes selon Aby Warburg* (Paris: Minuit, 2002); Giorgio Agamben, *Image et mémoire. Écrits sur l'image, la danse et le cinéma* (Paris: Desclée de Brouwer, 2004).

25. Aby Warburg, *Le Rituel du Serpent. Art et anthropologie. Récit d'un voyage en pays Pueblo* (Paris: Macula, 2003); Philippe-Alain Michaud, *Aby Warburg et l'image en mouvement, suivi de Aby Warburg, souvenirs d'un voyage en pays Pueblo, 1923, et Projet de voyage en Amérique* (Paris: Macula, 1998).

CHAPTER 3

PAGES WRITTEN IN EARTH

It gets you every time. The place is nondescript, with nothing to distinguish it from so many others just like it. There isn't anything special about it at all; not even its name is inspiring. You have been there barely half a day, and gotten no farther than removing the layer of topsoil, when large, dark spots appear, or perhaps the base of a wall running through the earth, or chalk-white skeletons scattered about. Whatever is lying there was just a foot down, no deeper than a high step. Some of the locals had been coming to this spot since their childhood, as had their parents and probably their parents' parents before them, totally unaware of what this site held. Up until this morning, you had been walking over these artifacts with no way of knowing they were present. And yet, they were there, so close, just below the surface, as if beneath the membrane of life. They all lay there, present and invisible, like the strange world of layers and walls interspersed with cavities, canals, and tendons that lie beneath our skin. And, in fact, within a few hours, just as a wound might, the stripped earth will have lost color and dried out, or else darkened and filled with water, depending on the terrain.

Like anatomy, archaeology shows us things we are not normally supposed to see. Some early 19th-century archaeologists were especially struck by the hidden similarity between archaeology and surgery. The English general Augustus Pitt Rivers (1827–1900), with a thoroughly military meticulousness, had the idea of assembling on Cranborne Chase, his property in southern England, models of the various sites he had excavated. For him, these were the culmination of his work, because they revealed the internal structures that excavation had brought to light, such as filled-in trenches and the inside of mounds raised for funerary monuments that dated back to the Neolithic period or the Bronze Age.[1] The models remind us of those wax anatomical figures that show how our internal organs are configured. You look at them and feel the same uneasiness you experience on a freshly worked dig site, for there is something intimate about both anatomy and archaeology (the inside of our bodies, the personal remains of people who once lived there), in spite of how generic they are: the organs that are exposed can be found in any human body, as can the interchangeable objects any period produces. Archaeological and anatomical meaning is a mystery, or rather it is not clearly established, because the disciplines are embodied in material creations.

There is nothing quite as powerful as the sense of intimacy, and nothing as short-lived. Once the life within has gone, once, as Proust put it, the "fragrance of melancholy, the enduring scent" of the past has completely evaporated, there remains just another empty vessel, just an ordinary carcass. Meaning has receded into its outline, but it remains there in a latent state, waiting for something to bring it back to life. It is as if it had been fossilized; its matter has given it a form that is, by its nature, peculiar. This explains why we regretfully turn away from these molded networks of veins that look like branches of coral, or from rows of ribs lined up like forks in a silverware chest, or from smashed skulls strewn about like so many broken eggshells.

The fact, then, is that we are eventually to be represented by a form, but we can't really recognize ourselves in this material object. Yet, we live with it. We carry our skeletons around with us the same way we live in our houses and handle utensils, without ever seeing them, even though they are the only part of us that will remain, the only authentic part of who we are as individuals and as mankind. These ambiguous, inconspicuous things carry our memory. They are here, among us, but we are blind to them. Actually, memory never surfaces by itself. It can remain in a latent state for thousands of years, and can just as easily never become manifest, much as artifacts might never be brought to the surface. It takes a chance happening for memory to appear. In his reading of *A La Recherche du temps perdu*, Walter Benjamin was taken with the passage in which Proust described the process of remembering which, he said, "bore the stamp of the very circumstances that had triggered it."[2] The passage is well known; it is the one about the *madeleines* at Combray. But of greater interest still are Proust's observations on the way the taste of the pastries from his childhood awakened his memory.

> And so it is with the past. It is a labor in vain to attempt to recapture it; all the efforts of our intellect must prove futile. The past is hidden somewhere outside the realm, beyond the reach of intellect, in some material object . . . . And it depends on chance whether or not we come upon this object before we ourselves must die.[3]

As archaeologists, we are invested with the peculiar ability to bring objects lodging memory back into the light, to extract them from the sediment of time in which they lie buried. We have learned how to locate stores of hidden artifacts or archaeological sites from concentrations of debris that are visible above ground. For the last thirty years, the widespread use of aerial photography has allowed us to detect the outlines of buried structures or trenches from the surface patterns of vegetation they produce.[4] Better still, with the progress in geophysical reconnaissance techniques that improved computer capacities have generated over the past ten years, we can now perform veritable underground endoscopies.[5] Differences in the electric conductivity or the magnetic intensity of given terrains can indicate the sites of walls, filled-in trenches, or hearths that are invisible above ground.

And yet, overall, the most impressive archaeological finds owe their discovery to chance. They serve to remind us, as Proust pointed out, that the past lies hidden in our material surroundings, more often than not in places we would never suspect. That is why most of these discoveries have been made by people not especially interested in the past, by people who tirelessly crisscross the land, or who are simply attentive to it. Pompeii was discovered by workers digging a well near the foot of Mount Vesuvius in 1689. They came upon stones with the lost city's name engraved upon them, buried beneath a large build-up of layers of lapilli. In China, in 1974, the discovery of the Terra-cotta Army near the gigantic tomb of Emperor Qin Shihangdi was also the result of well-digging. In Jordan, in 1947, the Dead Sea Scrolls were discovered by a shepherd looking for lost sheep. The stone he threw into a cave to drive an animal out happened to hit one of the jars containing the rolls of papyrus. Children have often been responsible for major discoveries. The Paleolithic cave paintings in Lascaux, in Dordogne, were found by four boys who had slipped down through a hole that had become visible after a tree fell. Earlier, in 1879, a little girl accompanying her father, Don Marcellino de Sautuola, as he explored the Altamira cave in Spain, came upon the first painted buffalo from the Upper Paleolithic. Looking up, she saw a herd of eighteen red and black buffalo running across the ceiling. As much out of fear as from excitement, she cried out: *Toros! Toros!*

We know that Freud was fascinated by antiques and archaeology. His office in Vienna overflowed with an eclectic collection of Egyptian, Roman, and Chinese figurines that he set up like an army marching toward him. Over the years, he completely covered the office walls with large display cases of artifacts and with engravings of archaeological monuments, like the Sphinx of Giza or the temples at Abu-Simbel. They hung over his couch, next to the bas-relief of the unreachable Gradiva. In all, there were 2,000 pieces in the collection he had begun putting together in the 1890s, after his father's death.[6] In 1938, the library in his office included some 2,500 works on archaeology.[7] Among them, prominently displayed, were Heinrich Schliemann's works, in particular *Illios* and *Mykenae*,[8] the monographs about his famous digs at Troy and Mycenae, and Howard Carter's writings on his sensational discovery of the tomb of Tutankhamun.[9]

Freud's vision of archaeology, which he drew from the great discoveries in the Mediterranean basin between the end of the 19th century and the 1930s, was dominated by what might be called, to borrow Lewis Binford's expression, "the Pompeii syndrome."[10] Like the mythical cities of Troy and Mycenae unearthed by Schliemann, the ruins of Pompeii discovered under the ashes of Vesuvius effectively give us to understand that creations of the past lie buried, intact, beneath a mass of subsequently deposited sterile layers that one has only to remove to unveil hidden creations as they were. Effectively, the digs at Pompeii and Herculanem, like those of the monuments in Egypt and Greece, were at first essentially large-scale clearing operations. The 18th-century engravings of the digs at Pompeii show the massive effort to remove volcanic waste. There are mounds of earth piled several yards high, closing the perspective of streets lined with Roman houses that are almost perfectly intact up to the second floor. We can see workers shoveling out the sterile soil and tossing it into wicker baskets that lines of women are carrying off into the distance, on their heads. Gradually, the ground of the ancient city is being brought to the light, with its high sidewalks and its streets paved with irregularly shaped large stones.

In Hissarlik, in the early 1870s, Schliemann employed teams of as many as one hundred fifty Turkish workers to dig nearly fifty feet down into the incredibly complex jumble of the superposed cities of Troy. From beneath several yards of soil and debris, there emerged the enormous stone ramparts that originally surrounded the citadel and whose base reached down into the bowels of the earth. The impact of Schliemann's digs has been even greater than that of the excavation of Pompeii, for his work speaks directly to the imagination. It stands as proof that mythical sites of great symbolic importance not only really existed, but that, in spite of thousands of years of oblivion, it is still possible for us to unearth them and bring them back whole. The great German scholar Rudolf Virchow, who visited the site at Hissarlik, was enthralled by Schliemann's digs. He watched "the city that had been 'burned to the ground' rise up whole from mounds of rubble accumulated over the centuries."[11] For Schliemann was absolutely certain that he had found Homer's Troy, just as he claimed to have found "Priam's treasure," which had been hastily assembled and then left behind by those taking flight.

A powerful archaeological metaphor, drawn initially from Schliemann's digs, was a direct source of inspiration for Freud. Like archeology, psychoanalysis is an exercise in resurrecting the past. As Freud confided to his good friend Fliess, with regard to one of his patient's recollections of a "scene dating back to his infancy (before he was twenty-two months old)": "It all happened as if Schliemann were once again unearthing what had been thought to be the imaginary city of Troy."[12] A crosscut of the great hill on which Troy stood—such as the one in Wilhelm Dörfeld's drawing, which Freud had in his library[13]—shows a series of layers of ancient cities superposed at regular intervals, and the faded outline of the last classical temples hovering above them, like the ghost of vanished remembrances floating above the sedimentation of memory. Like Schliemann, Freud dug down into what had been accumulated in memory. In reference to the case that first provided him with material he would later develop into his approach to psychoanalysis, he said: "It was my first full analysis of hysteria. It allowed me to use a method that I would later make into a technique, that of making my way down into the psyche, layer by layer, which we used to liken to the technique used in excavating a city."[14]

Actually, things are not quite as straightforward as Schliemann's massive clearing operations might lead one to believe. When dealing with complex sites, archaeologists quickly realize that foraging the ground to extract "antiquities" is not at all a simple affair, for the artifacts that they are trying to draw out have fused in part with the sediment in which they lie. As a rule, only a small fraction of the matter initially left in the earth remains. In most cases, the wood, leather, cloth, and various organic remains have completely disappeared. In tombs, the pottery set next to the body and the jewelry placed over it have generally been crushed by the decomposed walls of the coffin or funeral chamber that have come crashing down. Corrosion not only makes metal objects unrecognizable, it occasionally thins the metal so much that the objects become brittle or flimsy. At times, the level of acidity in the soil is so high that it can dissolve bones, those of a skeleton in a tomb, for example. In cases like these, only the arrangement of the objects originally placed on the body can give us an idea of how it was positioned. Often, it is just a particular coloring of the sediment that reveals their outline.

As we remove the sediment that has completely filled what were funeral chambers, or the earth from what were once dwellings, we constantly risk destroying artifacts, which are by their very nature short-lived, or of losing small pieces that have fused with the earth, or of breaking apart fragile fragments that it will be impossible to reassemble once the bits have been scattered. A dig thus requires the trained eye of experts who know when to be especially precise and careful. Above all, as they approach buried artifacts, they have to be able to anticipate their presence from certain signs visible on the surface, such as changes in the color or the texture of the sediment. This is why the earliest archaeologists from the end of the 19th and the beginning of the 20th century employed "master diggers" endowed with a sharp, intuitive sense of the land. Many of the major archaeological digs owe their success to these men, most of whose names are unknown to us.

And yet, very quickly, even these precautions prove to be inadequate. It is not enough to inflict as little damage as possible while unearthing objects or structures because their very setting holds information at least as important as that which we can glean from the objects themselves. Extracting artifacts from the ground and uncovering sites are basically exercises in destruction. We don't really bring to light a past that we had thought lost, as Schliemann believed he did. Rather, as we work, we precipitate the disintegration of the matter that holds the memory of the past. In effect, the search for objects on a dig quickly proves to be a destructive illusion. For a chance to enhance a collection with a few "beautiful pieces," we turn over a mass of accumulated archaeological layers housing debris that may well be devoid of esthetic value, but which nonetheless carries essential historical information about *context*; for example information with which to date artifacts. Thus it is that archaeologists themselves have, over the last few decades, come to consider amassing collections of ancient objects and uncovering ancient ruins, which once defined archaeological work, to be a process injurious to archaeological matter. It is a dilemma that still confronts us today.

Leroi-Gourhan and the "Ethnographic" Dig

Our awareness of the basically destructive nature of a dig has led us to seek techniques that are more attentive to the information about the past that archaeological sites contain. In the 1950s, the French prehistorian André Leroi-Gourhan developed a method he described as "ethnographic." Its purpose was to "uncover the very earth on which prehistoric men walked,"[15] in order to study the objects bearing witness to their practices and activities. With its emphasis on what remains of the ground, the technique would ultimately lead to a dig on a reindeer hunters' campsite in Pincevent (a hamlet that is part of La Grande-Paroisse, in the department of Seine-et-Marne) in the 1960s and 1970s.[16] For Leroi-Gourhan, the various surfaces on which Paleolithic men lived and which now lie buried beneath layers of earth, were akin to the pages of a book that had to be deciphered. In 1950, he presented his techniques and methods of prehistoric digs thus:

> You have to think of the earth as a book in which objects are the text . . . . And as a great prehistorian once observed, "it is a peculiarity of this book that it cannot be read without being destroyed." Any

> prehistorian who bores into the ground with the aim of enhancing his personal collection is like a man who would cut letters he especially likes, like O's and Z's, out of a precious manuscript and then claim, as he pastes them onto a sheet of paper, that he is engaged in scientific work.[17]

The fundamentally destructive nature of the archaeological dig, which "ends with the removal of objects from their context and the destruction of the document that was the unified whole"[18] literally obsessed Leroi-Gourhan. He considered the dig a "grave scientific act," with the archaeologist responsible not only for what happened to artifacts, but to the past itself from which these remains were exhumed. Information about the past is in itself evanescent and unreachable. It can be obtained only by "reading," or more precisely, decrypting the ground. As Leroi-Gourhan noted: "we have to save everything that bears witness to the past and that disappears under the prehistorian's spade" by concentrating on "what we can learn from the earth that lies around the collectibles," for "often, there is actually more written between objects than on the objects themselves."[19] The act of digging is thus more than simply reading those "pages written in earth" that are the ground archaeologists work in. As opposed to the letters on a white page that make up a written document, artifacts are not simply lying on the ground, and their meaning does not emerge from their unidirectional relation to each other. "It seems to me that it's no exaggeration," Leroi-Gourhan said, "to say that the ground is no less important than the objects we find in it . . . . The ground is actually better than a manuscript page whose text would be made of objects; it offers a great deal of commentary."[20]

How does one know what to retain from the seemingly undifferentiated matter that holds artifacts? How does one bring out the "commentary" that can give us the context in which these artifacts ended up here? There is no answer to these questions, at least no clear answer. And that is why, as Leroi-Gourhan emphasized, archaeologists have to record whatever fleeting traces they find in the ground before they vanish in the process of the dig. Whatever meaning these methodically collected observations have will appear in a very Freudian "deferred" phase. "It is to be expected," Leroi-Gourhan wrote, "that the smallest dig will generate considerable documentation and that just a fraction of that will prove useful, but only after the fact, and, more often than not, in unforeseeable ways."[21] For example, that loose grouping of tiny bits of flint and animal bones that seemed unimportant was actually there because it lay on the perimeter of an external hearth and tool cutting area that had not yet been uncovered. Or, those isolated, formless stone tool fragments that we didn't think would yield anything significant were in the ground there because they had been used to fill in a prehistoric pit farther down that we had not yet reached.

To work on a dig, you have to know how to "read" a space. And, as Georges Pérec pointed out, nothing is as problematic as a space. "The space of a life is neither continuous, nor infinite, nor homogeneous, nor isotropic . . . . The problem is not one of inventing a space, but of investigating it or, more simply still, reading it. For what we call daily life is not obvious; it is opaque, a form of blindness, a kind of anesthesia."[22]

Another Kind of Archaeology is Possible

At the beginning of the 20th century, the great German art historian Aby Warburg conceptualized art history as a "biology" of images, an anthropology of representations. He said he adopted this approach because he was "quite frankly disgusted" with "art history's insistence on esthetics" and the "sterility" of its debates and controversies.[23] Like conventional art history as seen by Warburg, archaeology also seems to direct a great deal of its energy to matters not directly related to its true purpose. The conventional approach to the archaeological past posits that history—past events, what really happened historically in the past—is the only tangible identity of the past that artifacts can express. It is widely believed that the past can be read from artifacts, and that this allows us to establish the reality of the past as it was; in other words, its historical identity.

But this is not at all so. There is no such thing as reading the past as it was. Its original identity was definitively lost in the process of fossilization through which it passed. Its material remains come down to us truncated, or enlarged, or transformed, and we are at a loss to distinguish what existed originally when the past was unfolding from the altered forms that evolved thereafter. The archaeological objects with which we work are essentially hybrids whose original matter and transformations are consubstantial. This means that what lies between us and the past, between the physical reality of the past as it originally existed and its archaeological transcription in the present, is far more than a filter modifying information about what it was initially, that it is far more than a bias we might adjust for. It means, in fact, that the very identity of the past, its ontological identity as it were, has irreversibly been changed in the time since it was buried because time has been working away at it ever since. Archaeology's grand vision is an illusion. There is no way to restore the past, no way to reconstruct it, for the past per se never existed. It exists only as it has been shaped by what came after it and gave it its existence, which is to say in a degenerated, deformed, and denatured state. We cannot move back from the present to the past by turning the pages of the "archives of the soil," to use the famous metaphor borrowed by Leroi-Gourhan. We can just reactivate a buried potential that can be dealt with only from the present, and that, more profoundly still, has no identity other than that which it assumes after-the-fact.

To borrow Husserl's terminology, we have to go back to the "things themselves," that is, to identifying artifacts at their source. In the 1980s, researchers from the American school of "New Archaeology" focused their attention on the phenomena involved in the way artifacts evolve. These "new archaeologists," anxious to break with the tradition of cultural history that had long been associated with their discipline, sought to root archaeological practices directly in the physical sciences. Along with Lewis Binford, Michael Schiffer was undoubtedly the person who looked most closely at the "processes by which archaeological matter is formed."[24] On the basis of concrete examples drawn from the present state of abandoned campsites, Schiffer methodically demonstrated that the nature of artifacts—what defines them as objects bearing witness to the past—was primarily the product of changes triggered by natural processes associated with the way in which they had come to be buried. As Schiffer noted, these degenerative processes, predominantly erosion and animal activity, directly affected the way in which the archaeological context of artifacts evolved.

In other words, artifacts first enter the ground not through human intervention, but as the result of natural causes that have disturbed or fragmented them. Schiffer's work was also noteworthy for having shown that even objects and materials discovered in the settings in which they had been used had been subject to constant manipulation: moved, reutilized, recycled, and so on. By studying objects that are a part of our present-day material culture, Schiffer demonstrated that they were never stationary, never stable, that they were rather continually denatured and, in a way, decontextualized. He showed that recycling and using things for unintended purposes was extremely frequent, and that various maintenance activities tended to disturb and relocate things, especially when it involved eliminating household waste that got carted off elsewhere. Last, in inhabited zones, in particular in heavily populated areas, he noted that buildings were constantly being renovated and taken over, just as abandoned structures are constantly being pillaged for construction materials. All of these practices have a direct effect on the artifacts that archaeologists eventually discover. They can dictate whether certain kinds of objects are present or not, as well as their size, number, disposition, and even their state of preservation.

Schiffer's work teaches us something absolutely essential about artifacts in that it deals with them in the present, where they are found. The "new archaeology" has been able to present a realistic picture of the phenomena involved in archaeological contexts because it has focused on the effect that the erosion of living spaces has on the materials of which they are composed. In effect, we do not know how the ancient archaeological formations that we interpret as "deposition layers" or " destruction layers" were actually formed. The only way to have some idea is to study how they are constituted in the present, with the aim of imagining how the end product—that is, the archaeological matter—emerged, by projecting onto the past mechanisms at work today. In a perfectly symmetrical way, the archaeology that deals with the processes by which artifacts are formed has been able to generate a cogent, or rather, an informed picture of the conditions under which objects were created in their past cultural context by examining the consequences of the way objects and materials are handled in present-day societies. As Schiffer noted, "the cultural past is knowable, but only when the nature of the evidence is thoroughly understood."[25] In other words, we cannot know the archaeological past as it was, in what was purportedly its initial state. We can know it only through the various transformations it has undergone, which is to say as matter subject to change.

Understanding that the space inhabited by the past is not the past itself but the present, and only the present, is a fundamental realization, one that very few historical disciplines have been able to come to, from their beginnings in the 19th century on. It is highly revealing that only those disciplines that deal with nature, the disciplines that distance themselves from humans, have been able to make this fundamental, epistemological leap that has allowed for the development of both geology, with Charles Lyell's[26] "modernist" approach, and paleontology, with Charles Darwin's revelation of the principle of natural evolution through reproduction.[27] Archaeology, and history in general, have never been able to take this seemingly suicidal step, thus condemning themselves to a history that is "closed," in other words, already written. Yet, as Lewis Binford noted, "The archaeological record is here with us in the present. It is out there, under the ground, quite likely to be hit by someone building a new road; it is very much part of our contemporary world and the observations we make about it are in the here and now, contemporary with ourselves . . . archaeology is not a field that can study the past *directly*, nor can it be one that merely involves discovery . . . it is a field wholly dependent on inference to the past from things found in the contemporary world."[28] Could it be that archaeology is, by definition almost, "archaeology of the present"?

Notes

1. Augustus Pitt Rivers, "Typological Museums," in *Journal of the Society of Arts*, no. 40, p. 115, December 18, 1891.

2. Walter Benjamin, *Œuvres*, trans. Rainer Rochliz and Pierre Rusch (Paris: Gallimard, 2000), tome III, pp. 333–4.

3. Marcel Proust, *Remembrance of Things Past* , trans. C. K. Scott Moncrieff and Terence Kilmartin (New York: Random House, 1981), pp. 47–8.

4. Roger Agache and Bruno Bréart, *Atlas d'archéologie aérienne de Picardie* (Amiens: Société des Antiquaires de Picardie, 1975).

5. Irwin Scollar, et al., *Archaeological Prospecting and Remote Sensing* (Cambridge: Cambridge University Press, 1990).

6. "I have sacrificed a great deal for my collection of Greek, Roman, and Egyptian antiquities," Freud wrote in a letter to Stefan Zweig, "have actually read more archaeology than psychology." Letter dated February 17, 1931, in *Letters of Sigmund Freud,* ed. Ernst Freud, trans. Elaine and William Robson-Scott (New York: Basic Books, 1960), p. 403.

7. Fred Botting and J. K. Davies, "Freud's Library and an Appendix of Texts related to Antiquities," in Lynn Gamwell and Richard Wells, *Sigmund Freud and*

Art. His Personal Collection of Antiquities (London and New York: Freud Museum and State University of New York, 1989).

8. Heinrich Schliemann, *Mykenae: Bericht über meine Forshungen und Endeckungen in Mykenae und Tiryns* (Leipzig: F. A. Brockhaus, 1878); Schliemann, *Ilios.*

9. Howard Carter and A. C. Mace, *The Tomb of Tut-Ankh-Amen. Discovered by the late Earl of Carnavaron and Howard Carter* (London: Cassel, 1923–1933).

10. Lewis Binford, "Behavioral Archaeology and the 'Pompeii Premise,'" *Journal of Anthropological Research*, 1981, 37, pp. 195–208.

11. Rudolf Virchow, introduction to Heinrich Schliemann, *Ilios, Stadt und Land der Trojaner. Forschungen une 57-Entdeckungen in her Troas und besonders auf der Baustelle Troja.* (Leipzig: F. A. Brockhaus, 1881), I.

12. Sigmund Freud, *Études sur l'hystérie*, trans. Anne Berman (Paris: Presses universitaires de France, 1956), p. 272.

13. Wilhelm Dörpfeld, *Troja und Ilion. Ergebnisse der Ausgrabungen in der vorhistorischen und historischen Schlicten von Ilion, 1870–1894* (Athens: Beck & Barth, 1902).

14. Sigmund Freud and Joseph Breuer, *Études sur l'hystérie*, trans. Anne Berman (Paris: Presses universitaires de France, 1956), p. 109.

15. André Leroi-Gourhan, *Les Fouilles préhistoriques (techniques et méthodes)* (Paris: A. et J. Picard, 1950), p. 13.

16. André Leroi-Gourhan and Michel Brézillon, *Fouilles de Pincevent. Essai d'analyse ethnographique d'un habitat magdalénien.* (Paris: CNRS, 1972), Supplement VII to Gallia Préhistoire.

17. Leroi-Gourhan, *Les Fouilles préhistoriques* 2. As early as 1862, the men who first entered the megalithic tumulus of Mont-Saint-Michel in Carnac (in the Morbihan department) said the following: "The past that remains here intact must be touched as one would touch something holy. For an encounter of this order, no precaution is too great, and it is important that the objects and the places be described as they were when they were discovered, for disturbing them accidently could wipe out in a second irrecoverable pages that had been respected over centuries." René Galles, *Rapport à M. le préfet du Morbihan sur les fouilles du Mont-Saint-Michel en Carnac faites en septembre 1862* (Vannes: Imprimerie Galles, 1862), p. 11.

18. Michel Brézillon, "La Fouille" in *Dictionnaire de la Préhistoire* (Paris: Larousse, 1969), p. 105.

19. Leroi-Gourhan, *Les Fouilles préhistoriques,* p. 3.

20. Leroi-Gourhan, *Les Fouilles préhistoriques,* p. 11.

21. Leroi-Gourhan, *Les Fouilles préhistoriques,* p. 4.

22. Georges Pérec, *Species of Spaces and Other Pieces,* trans. by John Sturrock. (London: Penguin Books, 1997).

CHAPTER 3

23. Ernst Hans Gombrich, *Aby Warburg. An Intellectual Biography.* (London: University of London, Warburg Institute, 1970), p. 88.

24. Michael Schiffer, *Formation Processes of the Archaeological Record* (Albuquerque: University of New Mexico Press, 1987).

25. Schiffer, *Formation Processes,* p. xx.

26. Charles Lyell, *Principles of Geology, Being an attempt to explain the former changes of the Earth's surface by reference to causes now in operation* (London: John Murray, 1830).

27. Charles Darwin, *On the Origin of Species by Means of Natural Selection, or The Preservation of Favoured Races in the Struggle for Life.* (London: John Murray, 1859).

28. Lewis Binford, *In Pursuit of the Past.* (London: Thames & Hudson, 1983), 19, 23.

CHAPTER 4

AN ARCHAEOLOGY OF THE PRESENT

In response to a journalist who had asked him how he had come to be a painter, Marc Chagall replied somewhat impishly, "Until I was about fifteen or sixteen, I didn't realize you could draw with just a pen. I thought you had to have a Masters in Fine Arts." I too thought that doing acceptable archaeology required some sort of accreditation. I thought I had to be a certified participant in the ongoing discussions among professionals. I thought that way for a long time, until I realized it was possible to be even more deeply engaged in archaeology simply by observing the things around me.

As a field of study, archaeology is a domain that begins right now, right here. It isn't necessary to seek it out on the other side of the planet, or buried deep down. It is here, within our reach. And what we need to be studying, with an archaeologist's eye, is the way the material remains that compose the heterogeneous mass of our present are interwoven. We can, for example, look at landscapes; urban and rural landscapes both have a great deal to tell us about our discipline. Or, we can look at living things, and all the built or manufactured products we use. Houses, objects, tools, and so on can all teach us much about the way archaeological time functions, how it is recorded in matter, and how it produces what we can only call history. We can look to see the connections between remains, separate the variable elements from the stable ones, and distinguish things that change from those that endure. The present, in this respect, opens onto all the pasts that have preceded present time and that are recorded in it. The archaeology of the present is in fact all of archaeology.

CHAPTER 4

An Archaeology of the Near Past?

So, we have to start from the present, that is, from the artifacts that constitute present reality. In practice, most archaeologists do not bother with such matters, because "modern" artifacts are, almost by definition, rejected as being "recent," which is to say "non-archaeological." Since archaeology purportedly deals with the past, it is taken for granted that its focus is necessarily on "ancient" things. The problem arises when a chronological line has to be drawn to separate what belongs to the province of archaeology from what does not. The issue has actually become a highly sensitive one, for in the course of the past fifty years the discipline's focus has widened to include recent eras. In France, medieval archaeology received recognition in the 1970s. Then, after much controversy, archaeology of the modern period (in a historical sense, that is, the 16th through 18th centuries) gained legitimacy in the 1990s, in the wake of the digs at the Grand Louvre. The question now before us is that of an archaeology of contemporary times (the 19th and 20th centuries).

As often happens in cases like these, the experts continue to debate the legitimacy of the discipline's expansion into new and seemingly incongruous areas even as these new areas are being worked and yielding results. In truth, once one begins systematically recording archaeological data, one sees that artifacts from the modern and contemporary periods are an integral part of older archaeological sites, and it generally becomes impossible to eliminate them as mere intrusions or simply the result of upheavals. In many instances, sites that developed during periods conventionally deemed "archaeological" (including most of the cities dating back to Roman times, and numerous religious structures from the High Middle Ages) have continued evolving well into periods that are generally considered non-archaeological because they are part of the present or, at least, of the recent past.

Standard archaeological practice needs to be rethought. The point is not that certain present-day sites have archaeological origins worthy of study, but rather that the transformations that these sites have undergone in the course of recent eras cannot be separated from their long-term history. The recent past, and the present as well, are just parts of a sequence that have naturally been added on to those that preceded them, either by augmenting sites or artifacts with new matter, or by transforming them through physical change. Fundamentally, these transformations are part of the long history of places whose origins usually date back to prehistorical times. Denying the archaeological identity of the present, the "contemporary past," is simply an esthetic or historical prejudice that has no basis in the nature of artifacts themselves. The present is fundamentally archaeological, in the same way that all the other periods of the past are archaeological, if only because the present is their historical extension.

Let's take this further. Are we saying that there already exist archaeological sites that are representative of this budding contemporary archaeology? In fact, there do exist such places, sites that bore witness to acts generally recognized as having marked the history of the 20th century and that have been preserved as they were in order to safeguard their material memory. Most of the attention has been given to the 20th-century's global-scale armed conflicts, in particular World War II. The Nazi extermination camp in Auschwitz, for example, is considered an essential material witness to the Holocaust. Similarly, the A-bomb dome in Hiroshima was added, also after much controversy, to UNESCO's list of World Heritage Sites, as testimony to the atomic apocalypse of 1945. In the Limousin region of France, the village of Oradour-sur-Glane has been preserved exactly as it was immediately after the war, to bear witness to the horror of Nazi barbarity. These places are not merely memorials commemorating recent historical events, they are also archaeological sites, in the full sense of the term, in that their preservation maintains, along with them, the material memory of the past to which they belong.

Oradour Refuses to Die

And that is exactly where the difficulty lies. Attempts to preserve sites that bear witness to recent history have made it clear, from an archaeological point of view, that such sites do not have a fixed place in the past. Fundamentally, their preservation is nothing more than a creative act aimed at securing the past that one wishes to commemorate in a discrete, which is to say fictitious, point in time. The case of Oradour-sur-Glane offers a particularly compelling example. The American historian Sarah Farmer studied the memorializing of the massacre carried out there in June 1944 by a battalion of the SS Das Reich division.[1]

The archaeological aspect of the preservation of this site is especially clear in that Pompeii was an explicit reference for the people who conceived the project in 1945. They wanted to show how life in Oradour had abruptly come to a halt on June 10, 1944, when almost the entire population of the village was exterminated by the Nazis. Here, as in Hiroshima and Auschwitz, it was crucial to put on view a place that had been frozen in time, fossilized by the historical cataclysm that had befallen it: a modern-day Pompeii, so to speak. It is highly revealing that no one at the time considered that, for the site itself, life would go on. *It* had not died in 1944. People went on driving down the road that passed through the village, farmers there still had barns where they housed their tractors and various other machines, families with gardens went on growing vegetables and raising animals, and so on. In a word, the physical life, the regular life of Oradour, carried on in spite of the devastating trauma that had ravaged the population. Consequently, in 1946, one of the principal measures taken in conjunction with declaring the site a national monument was to seal it off definitively by prohibiting all access to it. A barrier was erected around the village, and the entrances to buildings were boarded up while a new Oradour was built nearby. For Oradour to be preserved as a memorial, it had to cease existing as a place with human life, which is to say as an archaeological site. But with what consequences?

The administrators of the Office of National Monuments and Historical Sites who had been put in charge of the project failed to anticipate the amount of work needed to preserve the site once it had been frozen in time. Initially, they thought it would require only weeding and small masonry repairs that could be handled by county laborers. But, by 1946, they had to launch a veritable rehabilitation project, and the scale of the work kept increasing as the years went by. They had to shore up walls that had begun falling apart because of exposure to the elements, replace rotting wooden lintels with steel ones, repair stonework that was crumbling, and so on. And without the gardens that, whatever their drawbacks, had helped maintain the site, there was really no choice but to put in lawns to arrest the unchecked growth of parasitic brush and bushes. As time went on, and new parts continued to replace the original ones, Oradour took on the surrealistic look of a contemporary theme park based on ruins. Paradoxically, its fabricated appearance made it look more and more like Pompeii.

The preservation of Doctor Desourteaux's car is probably the most convincing evidence of the impossibility of locking archaeological structures in the past. On arriving in Oradour, the doctor was immediately taken off by the SS and killed. The vehicle, which has never been moved, was intended to remind us of the circumstances under which Dr. Desourteaux had to abandon it. The fact that the car on display belonged not to Dr. Desourteaux but to a member of his family is really of no import. What is most striking is that this object looks less and less like a car. In the years since 1944, the paint has peeled off, baring the rusted metal that has had to be coated with an anti-corrosive varnish. The smaller parts of the car, like the windshield wipers, the rear view mirrors, and the bumpers, disappeared long ago. The windows and the seat covers no longer exist either. The tires have literally disintegrated. The chassis now sits directly on the ground, slightly lopsided, and the bottoms of the car doors have fallen off. What we see today is not so much a car from the 1940s as an unrecognizable remnant from the past, or more exactly, a fossilized car from the first half of the 20th century as it appears to us in the beginning of the 21st.

The problems associated with preserving contemporary sites teach us something not easily understood about "classic" archaeological sites that to our mind belong to periods of history that have ended: Regardless of what we do, the past, as material creation, continues to exist and to evolve in the present. We are convinced that we have to maintain Auschwitz and Oradour *as they were* because we believe in a historical time that is linear and sequential, with each new time frame pushing back the one that preceded it. We fail to see that this is not the time of the artifacts themselves, and that in our desire to maintain these remains "in their past," we are actually encumbering them with additions that are foreign to them. We invent—obviously—a past for them that never existed, whose invasive mass ends up replacing the original past that dissolves and vanishes, never to reappear. In spite of all that we do, artifacts will deteriorate and age; that is their way of remaining with us, of existing in our present. Disintegration and disappearance play an integral role in the physical identity of sites and artifacts; they are their living memory in the present. Archaeological time does not stop when sites are abandoned. It continues to work away at the component matter, which is then assimilated into another environment where, imperceptibly, it holds the memory of other eras.

"It's hard to recognize," Simon Srebnik says in Claude Lanzmann's documentary *Shoah*, as he walks across a large, open field surrounded by trees, "but it was here. They burned people here." The only sound you can hear is that of his steps in the grass and birds chirping in a nearby forest. *Ya, das ist das Platz.* "Yes, this is the place. No one ever left here again . . . . It was always this peaceful here. Always. When they burned two thousand people—Jews—every day, it was just as peaceful. No one shouted. Everyone went about his work. It was silent. Peaceful. Just as it is now."[2]

The Return of the Dead

Oradour-sur-Glane is one example of an attempt to preserve a location immediately following the event that gave it new meaning as an archaeological site from World War II. In truth, we come upon contemporary artifacts quite regularly, in our everyday lives, although they generally go unnoticed as discarded trash. The last ten years, however, have witnessed digs of 20th-century sites that have been approached archaeologically. The earliest of them took place in 1990, with the exhumation of a mass grave that held the body of French author Alain-Fournier, who was killed in 1914 in Saint-Rémy-la-Calonne, in Lorraine.[3] During the 1990s, veritable programs of field research were organized, especially in northern France, around Arras and Amiens. I would like to look at one of these—the discovery of a British bomber in Fléville—for what it reveals about the way archaeology of the recent past relates to contemporary history.

In 1997, roadwork near the village of Fléville-devant-Nancy, in Lorraine, uncovered metal debris from an airplane. Remnants unearthed in a core sampling taken by the archaeologist Jean-Pierre Legendre were identified as belonging to a Lancaster Bomber, RA 502 Z, that had disappeared during the night of February 1–2, 1945 while returning from a mission over Ludwigshafen, in the Rhineland. In a crater driven some 14 feet deep, they discovered the engines, part of the fuselage with its British Browning .303 machine guns and, most significantly, parts of the crew's equipment: two parachutes, pieces of flight boots and strips of a flight suit with, intact in a pocket, an emergency medical kit including vials of injectable morphine, and a silk map of eastern France, Belgium, and western Germany. It was concluded that two men who had not been able to use their parachutes had perished in the crash. The archival research that identified the aircraft also yielded the names of three of the survivors of the crew of seven: William Anderson (bombardier), Victor Cassapi (flight mechanic), and Allan Jarmel (rear gunner). Asked for his reactions, Victor said the following:

> Locating the remains of our plane and learning how my buddies had died affected me deeply. For the last fifty-three years, I have thought about their deaths over and over. I have always felt the need to find some focal point that would allow me to honor their memory. Now there is one, and I feel they can finally rest in peace . . . Norman Tinsley's sister, Nora, and Andy James' sister, Gladys, have both said they feel infinitely better knowing what happened to their brothers, and how the French people in Lorraine have honored them. That's how I feel too, even though for so many years I avoided thinking about what I did and the things that took place during the war.

Vic's words represent the finest homage that can be given to archaeological work. As someone who had been witness to the events to which the remains discovered in Féville attested, Vic was not especially interested in the way archaeologists interpreted them. Their work was meaningful to him in that it brought back a past that had been thought to have vanished, whereas it simply lay buried underground. Vic afforded archaeologists an extraordinary role, far removed from that of experts in ancient history. For him, archaeologists were people who resurrected the past, thrust it into the present and, in so doing, altered history. As he stated, by discovering the physical circumstances of his comrades' deaths, archaeologists gave closure to a period of mourning that had remained open-ended because no one really knew what had happened. In fact, they did something that goes deeper still: They granted the past its place—Vic spoke of a "focal point"—in the present. And, having found anew its place in the present, the past, which had until then been floating, could at last have happened, and could bring solace in the here and now.

The objections to the new archaeology of the "recent past" are not to be dismissed out of hand. Critics find it pointless in that it offers nothing that isn't already abundantly documented in writings, maps, photographs, and films, which is quite true. But what is truly of interest in an archaeology of the contemporary past does not reside in the documentation of material culture. It is essential to consider the full implications, namely, that recognizing the legitimacy of an archaeology of the present, or the recent past, is an implicit challenge to the traditionally historicist approach to artifacts. In fact, it goes even further. In claiming that this "other" archaeology is fundamentally no different from the archaeology of other, more ancient periods of the history of mankind, one is effectively contending that the historicist's approach to archaeology is groundless, because the conventional approach holds that archaeological remains derive their meaning from their place in a linear, chronological sequence. And that requires knowing precisely which kinds of artifacts point to given moments in this chain, and on which combinations of types and styles this rhythmic unfolding of time that is peculiarly archaeological time is based. This is one of the fundamental assumptions of the traditional approach: Because the different typological and chronological sequences that archaeology pieces together all proceed from each other along a single chain, the supposed linearity of archaeological time is linked causally. In this way, the emphasis on causal relations dovetails with a form of historical explanation, which is effectively the point of chronological sequences based on the stylistic or typological evolution of artifacts. Unfortunately, this is not how archaeological time works. And, in so far as the archaeology of contemporary times deals with the recent past, it gives the lie to the spontaneous unidirectionality of this supposed historical time by thrusting the memory of the past into the present and reactivating a memory process that had until then lain dormant.

In effect, archaeology reactivates the past, or rather, to use Walter Benjamin's expression, it makes events "historical posthumously."[4] Strictly speaking, archaeology does not exhume parts of history that took place before and outside of it. Quite the contrary, it directly contributes to the construction of this history by inscribing them in the present. And that is what those who hold to historicist orthodoxy find scandalous. In allowing that certain past elements can enlighten us about the present, this "archaeology of the present" blows the locks off time in that, as Walter

Benjamin noted, it acknowledges this capability with total disregard for the causal chain linking one moment to the next, which for the historicist represents the basic structure of historical time. It doesn't take it into account at all, because past events can be reactivated after a period of latency lasting years, or as Walter Benjamin said, "millennia."

We know this to be perfectly true. In the Epirus region of Greece, work carried out under the direction of British prehistorian Geoff Bailey showed that the severe erosion triggered by the excessive numbers of sheep and goats that had been brought into a sensitive area had reactivated a process of soil degradation that had actually begun in the Paleolithic period.[5] Bailey effectively demonstrated that the reconstruction of history, in so far as it recognizes links to the past such as these, can no longer pretend to function as reconstitution, as had been accepted until then; it rather functioned as transcription. In the same way, the importance we attribute to past events lies less in their function as "primary causes" than as the points in time that we recognize them to be after the fact, and that give meaning to this relational history.

It is the process of memory that dissolves the conventional time of history. For an archaeology functioning as memory in which time no longer plays its habitual role, the regular tick-tock of one period after another ceases to be central. We are led to deal with the whole of the present, through which all of history is constantly being re-evaluated. As Walter Benjamin said, it is the present as "nowness" (*Jetztkeit*) that, in place of the sequential, linear vision of the historicist, is the very source of history, the origins of which are endlessly "at present." "History," he said, in Thesis XIV of "On the Concept of History" "is the subject of a construction whose site is not homogeneous, empty time, [historicists' time] but time that is filled full by nowness."[6] As archaeology demonstrates, there lie within this unsettled present fragments of times gone by. When they are identified, they allow history to emerge. And this is really what Vic Cassapi was saying. Given that the "flashes" he referred to, which suddenly allow us to bring together the present we experience and a past revealed to us anew, trigger something truly miraculous that transforms history, it is hardly an exaggeration to characterize them as "messianic."

CHAPTER 4

From an Archaeology of Landscapes to an Archaeology of Memory

Studying the way in which material remains are lodged in the heterogeneous mass of our present is exactly what Gérard Chouquer did in analyzing landscapes, which are also archaeological structures.[7] Over here, we have an urban-scape, a gray and white city stretching from a petrified river basin to the tops of the surrounding hills, and beyond to the horizon; and over here, barren countryside, like a huge blanket of green and yellow fields thrown over the land; and so on. What is the history of these landscapes, and how does one gain access to it? We are no longer dealing with an imagined space from some specific past; nor are we bringing back into the light, at some minuscule point on the vast canvas that stretches over the entire expanse, artifacts from some vanished era about which we can say: "This is the past; this is how it was." We are here, in the present, looking out over a landscape that lies before us, a landscape that holds the past, all the moments of the past mixed together. We are looking at *form*, examining a surface crossed with lines and motifs, recognizing shapes, discerning the effects of an infinite variety of proximate forms that combine with each other to create overall patterns. One looks at a landscape as one looks at a painting, because there is nothing else to see other than pure form and morphological pattern. All the rest of it—all the supposed proof that the past existed in one fixed state or, more precisely, a chain of distinct states running throughout time—has evaporated as if it had never existed. Everything we see is permeated with the harmonious strangeness of the place, prompting new questions. What in this landscape changes and what remains the same when our perspective moves from close up and adjacent to distant and overall? How can this homogeneous space, or in any case this manifestly structured space, incorporate all that is heterogeneous, random, and foreign to it?

Questions about space immediately lead to questions about time. Here, time and space are fused, precisely because the past has not been buried beneath the present like some souvenir. It is present in matter, as a feature, and as part of the design and the patterns of the space. Take that city, for example, surrounded by industrial parks and suburban housing developments, or those vast landscapes carved out by industrial-scale agriculture; the structures that date back to ancient times—the roads and buildings and the networks of fields—have long since disappeared from the face of the earth. There is not a single visible trace of them, at most a few minute pieces of debris buried in the ground, and yet . . . . And yet, these works of the past still exist in boundaries, orientations, contour, and mass. They still delimit the physical structure and the organization of the space in the present. And no one particular moment of the past takes precedence over others, even if the imprint left by certain historical periods (that of the Roman occupation, especially) is particularly strong. It is rather the accumulation of all of the time frames of the past which, working together, not one after the other, configures the landscape. Here, the accumulated works of the past function not through simple, mechanical superimposition, but in a complex dynamic in which objects from a distant past continue to have an impact. And it is because archaeological events are inscribed in matter that their impact reaches beyond the point when they occurred. In fact, their impact will last as long as those archaeological events remain inscribed in matter.

Our concept of time has been challenged once again, because we are considering it from the present, and not from some imagined vantage point that would place us outside of its movement, and allow us to watch the procession of connected sequences pass by one after the other like a carnival of History on parade. At bottom, the search for the past is nothing more than the study of the material present. In the same way, the study of the past can be meaningful only in the overall perspective of all moments of time, not in the examination of one of them in particular. The memory held by places and things that develops in a time that is not the sequential, linear time of conventional history is the subject of archaeology. We need new words to describe the connections across time that objects from the past establish with each other. Beyond the conventional concepts of synchronicity (existing at the same moment, whether similar or different) and diachroneity (what is maintained in time) we need to include, as Gérard Chouquer pointed out, the more complex notions of uchronicity and hystereochronicity, the former designating the indeterminateness that characterizes every moment of time with regard to its future (that is, to the forms it might eventually assume), the latter designating the period of latency frequent in the history of sites and human dwellings, between an event occurring at a given moment and its subsequent effects, which are sometimes felt a great time later.

In his approach to the archaeology of landscapes, Gérard Chouquer emphasized a notion borrowed from ecology—*morphological hysteresis*—to designate the period of latency that becomes apparent "when the effects of a phenomenon that has died out are either felt long after or actually continue to develop through inertial forces."[8] And we could add to this list of temporal phenomena at work in archaeological memory the notion of prochronicity that Blaise Cendrars made use of in his fictionalized memoirs published under the title *L'Homme foudroyé* (*The Astonished Man*). In Cendrars' work, prochronicity appears as the ability of certain events to communicate across time and space, suddenly to reactivate certain past events, making them truly prophetic. Cendrars' rebirth as a writer was triggered by the memory that came back to him in 1940 of the circumstances under which his hand was blown off during World War I.[9] The dynamics of "polycyclical" archaeological processes are not that different. The successive making, sealing, and reactivation of archaeological constructs occur within cycles that follow each other at irregular intervals. For example, more than a millennium after having been abandoned, the sites of funerary monuments of the Neolithic might have come back as burial grounds during the Iron Age and then, nearly a thousand years later, as the site of a cemetery during the Early Middle Ages.[10]

These are interesting lines of inquiry, but in themselves they are not enough to lead anywhere. We know now that the temporal connections between archaeological constructs are not the "usual" ones, that they develop with a certain kind of memory that is not—strictly speaking—historical memory, but an internal memory, one that is intrinsic, as it were, to archaeological matter. If we are to give ourselves the means by which to understand the temporal interplay, we will have to focus on the mechanisms at work in the memory of archaeological constructs. It is pointless to return to the well-known, comforting world of traditional archaeology. In fact, there is no turning back; we have come too far. This is new terrain, and we have neither a name nor a map for it. It is a kind of archaeological memory that, curiously, resonates with psychoanalytic memory, not because they are both matters of the psyche (one collective, one individual), but because they are both rooted in matter subject to the effects of time and irreversibility. As the psychoanalyst André Green said, "Memory is . . . a multiple system of traces which were re-inscribed periodically, being 'retranslated' in the light of new circumstances."[11]

Traces, vestiges of the past organized in *systems*, *periodic* re-activation, the *re-inscription* in matter of archaeological objects that translate into stratigraphic superimposition, the reworking of archaeological structures from the past that allow them to *translate* otherwise in the present—we are, paradoxically, on familiar ground, that of the similar behaviors of artifacts and our memory, which is to say the workings of the unconscious. This is not to say that the two are identical, but that they are constructed in much the same manner. In both cases, remembering, or if you prefer, the resurgence of the past, does not proceed by drawing on some supposed stock of memories, which, for archaeology, would mean a store of ancient artifacts. It happens in the present, through the *re-inscription* of structures inherited from the past that have translated differently. To take an example at hand, a city's identity is maintained through the daily practice of redevelopment, reconstruction, and growth, not by freezing its downtown, if it were even possible to preserve it "as it was." This would only turn this "memory of the past" into a museum, or ruins, in any event into something that would in no way resemble what we understand by a city today.

Thus, the identity of a site that comes down to us, which is to say the memory it holds, is essentially perpetuated as it has been "translated" in the present, as it has had to adapt continuously to the new demands placed on it by current use. I live in a house that dates from the 17th century. I maintain its identity as a dwelling by continuing to live there, which is to say by transforming it to meet my needs. In doing so, I clearly undo its resemblance to the original, but that is not important. What is important is that it continues to function as a house. In effect, the changes I make become an integral part of it. This kind of history, where things are remodeled, recycled, and created, is of foremost interest to archaeologists. André Green stated it clearly: "Psychoanalysis is not very concerned with remembering; its real object is temporality."[12] Like its counterpart, archaeology too has but a secondary interest in moments of the past that can be exhumed in the present. Rather, both disciplines are fundamentally engaged in studying the way the identities of things and beings are transformed by their own history. The question that unites them is one of knowing, basically, the workings of time. In other words, what does time reveal?

Time's Subconscious

Having arrived at this point, the question now is how to take this further. The idea of delving more deeply into the internal memory of archaeological systems (that of groupings of objects, of sites, or of complexes of sites) presupposes for these systems the existence of an intrinsic identity that continues over time. This is not the identity attributed to them by historians, ethnologists, and archaeologists, or even by the people in the past who created or introduced the things whose vestiges we find today. The identity in question is one that is developed by and for archaeological systems as shaped by their own history, their own trajectory in time. Still, relatively speaking, this too is of secondary importance. The essential problem is that of determining the processes through which this memory is generated.

What does psychoanalysis have to say about this? In his work on time and memory, André Green noted that when something is assimilated as memory "the new registrations of earlier systems of traces are carried out not, . . . in order to introduce change but to avoid it."[13] This is an essential point, and once again one that is shared by archaeological memory. The perpetuation of archaeological manifestations is based on a complementary relationship between structures already in place and those that are subsequently added to them. Prehistoric funerary sites offer numerous examples of this phenomenon. When new kinds of monuments to the dead (such as the quadrangular enclosures from Gaulish times) are added to older memorial structures (the circular tumuli of the Bronze Age, in particular), they tend to be oriented along lines predetermined by the earlier structures. In certain cases, they even become part of the original monuments, a kind of annex, albeit one that had never been envisioned in the original design. In periods of recorded history, the changes undergone by churches and shrines—which, moreover, are very often situated on sites of worship dating back to Antiquity or Proto-history—offer innumerable examples of a similar kind.

As I have already noted here, the process of adding on to archaeological constructs fashions the memory they hold and extends their identity over time. But to be incorporated, innovations have to find their place, somewhere between novelty and tradition, within the environment that is to receive them. If what is new is overly incongruous, it risks being rejected, which means that subsequently it will prove incapable of being reproduced or generating an archaeological lineage. Conversely, if what is new is not sufficiently novel—for example, if it is more or less just a copy of past constructs—it will simply be absorbed within the preexisting archaeological context and will not be recognized as new. Whatever the case, incorporating new elements is vital to an archaeological system, for absent evolution and transformation, it will inevitably break down under the imperious pressure of the present, which relentlessly introduces new things. Thus, an existing system of archaeological constructs has to absorb innovations. And what is new is actually absorbed very well—at the same time that the change is all the more apparent—because it can be absorbed only in so far as it corresponds to the norms of the system already in place. And this, according to André Green, is how the system protects itself against "the threat of disappearance represented by what is new . . . , for the system is a depository not only of the past, but of *the pre-forming organization* of the present."[14]

Thus, beyond the adaptive process, the memory at work in the development of archaeological structures involves an organizational process. Clearly, as they evolve, something is constantly being repeated, something that regenerates previous forms and works to maintain the particular way that places and objects function and are organized. Whatever remodeling and renovation a house undergoes, what is carried forward is still a domestic living space, just as a cemetery remains a cemetery in spite of new graves and regardless of the kinds of tombstones planted there. In its various reaffirmations of the archaeological organization of structures, repetition appears as a response to change. When a house or a neighborhood is redesigned, it is because of new needs and demands that require that they adapt to this change.

Whereas the identity of archaeological sites remains, even after changes occur, their disposition is altered. These altered dispositions, which appear in physical contexts that are different every time, are evidence of the various functional identities that archaeological structures can assume in the course of their successive transformations; and the changes that occur suggest as well a variety of other transformations that the system in question could have undergone. It is clear that at every moment of a form's evolution something could, not just appear, but take hold and become permanent, ultimately leaving its mark on the archaeological system's evolutionary path. This is how Freud put it in a well-known letter to his doctor friend Wilhelm Fliess:

> As you know, I am working on the assumption that our psychic mechanism has come into being by a process of stratification: the material present in the form of memory traces being subjected from time to time to a *rearrangement* in accordance with fresh circumstances—to a *retranscription*. Thus what is essentially new about my theory is the thesis that memory is present not once but several times over, that it is laid down in various kinds of indications.[15]

As Freud noted, the "stratification process" (whose concept, as we have seen, he clearly borrowed from the archaeology of his day) does not really function as layers superimposed on each other. It more closely resembles the phenomenon of "reprising," or retranscribing the series of states that preceded the new rendering of the given archaeological manifestation. Digging a new network of ditches along the same lines of the earlier boundaries of some space does not simply superimpose—at times, after long periods of abandonment, after a longer period of latency—a new archaeological structure on an older one. It reactivates the older structuring of the space and, in so doing, gives it a new existence. Studying these superimpositions is, in this way, vital to our understanding of the transformations of archaeological memory.

As Freud pointed out, the "deposited" memories that compose the successive states of a palimpsest laid down in various kinds of "evidence" (which, for archaeologists, would mean different categories of deposits) are proof that memory is not single but several, which is to say that it is transcribed each time in a discernibly different way. It is not really the same site that is being redeveloped, but structures whose time frames and environments differ, even if they share, despite their diversity, a certain common identity that is being reactivated and reaffirmed. The study of the centuries-long additions to, and remodeling of, settlements and cemeteries clearly shows that the forms in which sites are preserved can vary considerably over time. And this stands as proof of the process of repetition that is at work in constructing and maintaining archaeological structures. When we look at a mass of similar artifacts—for example, hundreds of tombs gathered in a cemetery—we are actually seeing the repetition of a burial arrangement whereby each new grave is defined not so much in respect to other contemporary graves as to the aggregate of all of the dead people, be they known or not, who are already entombed there. In turn, with each new body added to the cemetery, its identity is at once maintained and modified. At bottom, innovation and change are introduced through repetition, which is what allows what is new to be inscribed in what came before; in other words, in this "past in the present" that defines the things around us.

The Ephemeral

How is it that archaeological systems, both those of objects and of sites, are maintained through repetition? Why must individual events be reproduced for the identity of archaeological constructs to be preserved over time? The simple answer is that repetition is a fundamental part of the memory process, of the sequential recording of the past: For there to be memory, there has to be reiteration, which is to say stratification. But, in itself, that answer is not sufficient. It helps us understand more clearly how archaeological palimpsests are fashioned—through repetition and retranscription—but it doesn't tell us why things proceed this way, and not otherwise. The key to the workings of archaeological memory lies in the nature of archaeological matter. In effect, repetition is the consequence of the ephemeral character of the things, both organic and non-organic, that humans create. In this world, nothing lasts forever; in fact, as we have just seen, nothing even remains as it was. Seemingly immutable artifacts are in truth hard pressed to hide their basic transience. Their very number attests to the instability generated by the archaeological process of repetition. Faced with perpetual deterioration, the function and identity of an archaeological construct have to be constantly reaffirmed and readapted; faced with the pressures exerted by a continuously evolving present, repetition is the *sine qua non* of perpetuation: inhabited places—dwellings, cemeteries, work spaces, etc.—owe their existence to the repetition of individual acts and transformations that allow them to remain functional. Once abandoned, they die, the reproductive work that erodes and eventually deconstructs them having been definitively interrupted.

This is why all things archaeological, be they artifacts or stratified sites, can fundamentally be likened to a palimpsest, the layered accumulation of memory. The need for repetition reveals the paradoxical link between the memory and the impermanence of beings and things. Objects have to be constantly (re)produced precisely because they are ephemeral, and it is through repetition that the conditions for the generation of the internal memory of archaeological constructs are created. In the same way, it is because the individuals who constitute communities have limited life spans that the duplication of tombs allows their memory and history to be recorded. It is because matter disintegrates and falls apart that it has to be constantly reworked and transformed, and its history is recorded in this newer form. And finally, it is because the present engenders the dismemberment and disappearance of all that is, that every act, every series of operations that transforms matter, is a new act and a new procedure. And in this endless process lies the promise of history.

Archaeology teaches us this paradoxical lesson about ourselves and about history: It is because we forget that we can remember, because we drop like flies that we can perpetuate ourselves. Of course, here, one has to understand by memory not something that once existed being recalled, but something as it is reinvented, just as one has to understand that perpetuating the past does not mean preserving it, but transforming it. The contents of the "black box" that was handed down to me by my mother and that I mentioned earlier in this book is an excellent example: As a reliquary, the objects included in that collection invite the addition of new ones which, because they will be different from those already there, are destined to change their identity irremediably. What is fascinating about archaeological images lies precisely in this paradox: Archaeology exhumes ephemeral witnesses to a past that disappeared while recording history, and these witnesses to the past have become the matter of ancient history only because they were ephemeral. Only one other medium presents this phenomenon: photography. As Roland Barthes said in *The Light Room*:

> The *noeme* of photography is simple, banal; no depth: "*that has been.*" I know our critics: "What! a whole book (even a short one) to discover something I know at first glance?" Yes, but such evidence can be a sibling of madness. The Photograph is an extended, loaded evidence—as if it caricatured, not the figure of what it represents (quite the converse),

> but its very existence. The image, says phenomenology, is an object-as-nothing. Now, in the Photograph, what I posit is not only the absence of the object; it is also, by one and the same movement, on equal terms, the fact that this object has indeed existed and that it has been there where I see it. Here is where the madness is, for until this day no representation could assure me of the past of a thing except by intermediaries; but with the Photograph, my certainty is immediate: no one in the world can undeceive me.[16]

The archeological image of our evanescence is visible in photographs of the mass graves in Srebrenica and Vukovar taken by Gilles Peress in 1997 for the International Criminal Tribunal established for the former Yugoslavia.[17] There is, in these pictures, enough to drive one mad, for the bodies piled in common graves show us the archaeological truth of who we are: barely human faces emerge from the earth, the enamel on their teeth conspicuous in a slick, gritty, whitish mass of skin, and neither eyes, nor nose, nor lips, because the digestive processes of the soil have absorbed them. We note a disproportionate amount of cloth, of tattered rags like so many elegant folds crushed flat against formless corpses, and an odd array of objects that have remained perfectly intact: a white sneaker and a tightly fitting sock at the lower end of leg bones, the plastic soles of cheap shoes humbly lying there side by side, a very ordinary metal watch that still reads 2:45.

This collection of tattered rags with loose folds is very much like the ones that were discovered when the Nazi extermination camps were liberated. Rags hang there, in the midst of heaps of fleshless, discolored corpses, or droop down from mountains of clothes piled high.[18] One finds the same incongruous objects, tenacious in their indestructible triviality, like the enamel-coated plates strewn over the abandoned railroad tracks that lead into the deserted, snow-covered camp at Birkenau; the same corpses lying on top of each other in a jumble of arms and legs that looks like a pile of cut branches (the survivors of the camps reported that when the gas chambers were opened bodies "fell out like blocks of stone, like rocks falling out of a truck"[19]). We note the way shoes tend to lie next to each other on their sides, showing their soles. This is the archaeology of our century: a snapshot of ephemeral daily life suddenly frozen in time, suddenly become a physical vestige while the human elements—the bodies and the people—fade and melt into indistinct matter. This is what captures the human state, what fossilizes our transience. This is what we are.

The result, as Benjamin pointed out, is that history is read in "rags" and "scraps of clothing." The historian who throws off the historicist shackles is, at bottom, just a "ragman" (*Lumpensammler*), a man who picks up the tattered shreds of the past. He rummages through the ruins of history looking for refuse, for ephemeral remains that lie crushed beneath the things that have collapsed beneath the rubble and the fill. The agenda for this new historical materialism (which, as the reader will have surmised, is the same as that of the archaeology to come) explicitly requires us, as Benjamin put it, "to create history with the very detritus of history."[20] This approach is irreconcilable with the traditionally historicist one that seeks, as we have seen, to untwist the pluri-linearity of artifacts. Searching for "scraps," a "rag-picker of history": It is remarkable to find exactly these terms in the approach to archaeology first delineated by the great antique collector Caylus, in the middle of the 18th century. "Please bear in mind," he wrote to Father Paciaudi in 1758, "that I am not putting together a display, that self-satisfaction is not the point, that I care not at all for show pieces, that the objects I seek are rather the 'scraps' of agate, stone, bronze, earth, and glass that might somehow help me learn about how things were used or the person who created them."[21] A short time later, writing to the same man, he said: "I liken antiques to handsome gentlemen accompanied by perfectly arrayed, beautiful women who arrive at some gathering, show themselves, and from whom we learn nothing, whereas from some rough piece, which I compare in this instance to a grimy fellow arriving on foot, I can occasionally draw matter for a treatise or a discovery . . . . In that, I am much like a 'rag-picker.'"[22] In the 19th century, Sigmund Freud would define psychoanalysis as the study of the "detritus" in an individual's psyche. At bottom, the two are one and the same.

Scraps of the present: that is, in its essence, what photography captures. *Smoke* is American photographer Michael Ackerman's 1999 collection of pictures of Cabbagetown. His portrait of the inhabitants of this destitute Atlanta suburb presents them as lost and suspended in time. By bringing out the striking forms their bodies have assumed in "nowness," the pictures capture the intensely transient character of these people's existence. "My photographs," Ackerman said, "tell the story of time going by . . . I tried to capture what went unspoken, and perhaps their ghosts . . . . Everything disappears immediately, like smoke." Ackerman has since published two other remarkable albums: *End Time City*, shot principally in Benares in India, and *Fiction*, a collection of shots of various European cities.[23] His photography has become subtler, simpler. It is hard to make out the shadowy, almost insubstantial faces, much like those in Srebrenica and Vukovar. The light play over their bodies produces dark spots, lines, and smoke-like wisps. Here and there in the vacillating images, one sees the imprint of teeth or rows of ribs that show through the outer membrane of the bodies in the mass graves and the cloth into which they disappear.

Ephemerality reveals the momentarily discernible vibrations of time. A lot could be learned from an archaeology of the present, whose purpose would be precisely that of observing that part of the present that is ephemeral: that which comes back, always the same and always different; that which does not endure, but persists. Here, the present is very much an intermediary space, an "in-between-world," to use Paul Klee's term.

> Worlds have opened and are constantly opening to us, worlds that also belong to nature, but that are not visible to everyone, that may only be visible to children, madmen, and primitives. I have in mind the world of those who are not yet born and of those who have died, the world that might come, that yearns to come, but that may not necessarily come: an intermediary world, an in-between-world.[24]

And yet, there is something inscribed in this space, this "between-world" that is the present. Something peculiar to it takes shape here and now. What is, then, this present that we inhabit?

Notes

1. Sarah Farmer, *Oradour. Arrêt sur mémoire*, trans. by Pierre Guglielmina (Paris: Calmann-Lévy, 1994).

2. Claude Lanzmann, *Shoah, An Oral History of the Holocaust* (New York: Pantheon Books, 1985), pp. 5–6.

3. Frédéric Adam, *Alain-Fournier et ses compagnons d'armes. Une archéologie de la Grande Guerre* (Metz: Serpenoise, 2006).

4. Walter Benjamin, *Selected Writings*, "On The Concept of History," Thesis A, vol. 4, p. 397.

5. Geoff Bailey, *Palimpsests, Time-scales, and the Durational Present: Origins and Consequences of a Time-Perspective Research Agenda* (University of Newcastle, unpublished manuscript, 2003).

6. Benjamin, *Selected Writings,* "On The Concept of History," Thesis XIV, vol. 4, p. 395.

7. Chouquer, *L'Étude des paysages.* His research drew upon the work of the geographer, Augustin Berque, who developed the concept of *écoumène* to describe human environments. Augustin Berque, *Écoumène. Introduction à l'étude des milieux humains* (Paris: Belin, 2000).

8. Chouquer, *L'Étude des paysages,* p. 150.

9. Blaise Cendrars, *L'Homme foudroyé. Le Sans-Nom* (Paris: Denoël, 2002).

10. For example, the Quedlinburg tumulus in northern Germany. An analogous phenomenon has been observed with Bronze Age tumuli on the Calotterie site in Picardy (Richard Bradley, *The Past in Prehistoric Societies.* London: Routledge, 2002, p. 128 and fig. 5.6, pp. 132–3, and fig. 5.10).

11. André Green, *Diachrony in Psychoanalysis, trans. Andrew Weller* (London, New York: Free Association Books, 2003), p. 182.

12. Green, *Diachrony,* p. 146.

13. Green, *Diachrony,* p. 183.

14. Green, *Diachrony,* p. 183.

15. Freud, letter dated December 6, 1896, in Sigmund Freud, *The Complete Letters of Sigmund Freud to Wilhelm Fliess, 1877–1904,* trans. by Jeffrey Moussaieff Masson. (Cambridge, MA: The Belknap Press of Harvard University Press, 1985), p. 207.

16. Roland Barthes, *Camera Lucida,* trans. Richard Howard (New York: Hill and Wang, 1982) p. 115.

17. Éric Strover and Gilles Peress, *Les Tombes. Srebrenica et Vukovar* (Zurich, Berlin, New York: Scalo, 1998).

18. Georges Didi-Huberman, *Ninfa Moderna. Essai sur le drapé tombé* (Paris: Gallimard, 2002) fig. 54.

19. Lanzmann, *Shoah,* p. 125.

20. Walter Benjamin, *The Arcades Project.* (Cambridge, MA: The Belknap Press of Harvard University Press, 1999), p. 545, quotation from Rémy de Gourmont.

21. Father Paciaudi, letter II, dated February 12, 1758, in Charles Nisard, *Correspondance inédited du comte de Caylus avec le P. Paciaudi, théatin* 1757–1765 (Paris: Imprimerie nationale, 1877), p. 4.

22. Father Paciaudi, letter III, dated August, 28, 1758, in Charles Nisard, *Correspondance inédited du comte de Caylus avec le P. Paciaudi, théatin* (1757–1765). (Paris: Imprimerie nationale, 1877), pp. 8–10.

23. Michael Ackerman, *End Time City* (Paris: Nathan-Delpire, 1999); *Fiction* (Paris: Delpire, 2001).

24. Paul Klee *(Souvenirs) Diaries,* quoted in Jean-François Lyotard, *Discours, Figures* (Paris: Klincksieck, 1971), p. 224.

CHAPTER 5

A FIELD OF RUINS

The End of History

On the TV screen, the air seemed saturated with dust. Frightened figures as white as ghosts were trampling countless sheets of paper that littered the streets. It was hard to make out what was happening. People were running away through clouds of smoke and ash that darkened everything. In an instant, time's movement forward had been halted, the present thrust into the past, beyond our reach. All that was left was an incredibly huge pile of incomprehensible and vaguely poignant vestiges—the stuff of archaeology. There now lay exposed the true present of things and of matter: a mix of incongruous bits of the past thrown together in a now that had been brought to a standstill. Of that perpetually new world that had led some to declare the End of History, there remained only rubble, junk, and filth.[1] Emergency crews were already busy cleaning it up, plugging this unimaginable breach in reality's protective shield. There would be no way to recover the thousands of crushed and burned bodies that had been reduced to shreds of cloth and flesh. They were inextricably mixed in with the hunks of concrete and steel that were being carted off to the dump on trucks—like rubbish, from which they were now indistinguishable. We know that disasters like the one at the World Trade Center can happen any time, anywhere.

The "short 20th century" witnessed the foundering of all of the 19th-century European ideals that had grown out of the Age of Enlightenment.[2] With the invention of new processes for the "industrial scale manufacture of corpses," to use Heidegger's phrase, the century's major events not only shattered man's belief in the beneficial effects of shared technological progress, universal education, and the preeminence of reason; they definitively rendered these tenets indefensible. History has not, as Fukuyama claimed, reached its culminating point at the end of a long journey supposedly begun with the appearance of mankind; rather, the past has become unintelligible, and the promise of progress that the future supposedly held has fallen away. We are now faced with a new experience of material reality, one manufactured by machines in quantities never before seen. *There is no Future*: We find ourselves cast into a world of millions of mass-produced objects designed to be discarded once they have been used, and then sorted, stocked, and finally (re)processed. Like them, we won't die; we'll disappear. The elimination of waste is just one particular kind of *reprocessing*, in that its purpose is not the production but rather the disappearance of goods and people. The rhythm of our lives is now that of the real time of a world in which goods circulate globally. It is a generalized present that is always new, with neither a "before" nor an "after." Reality is what happens here and now, at this moment, and nothing—not even we ourselves—has an existence prior to or following this absolute now. The transformation of the world has affected not only the individual's relation to his human and material surroundings, it has changed in a more fundamental way what defines the living beings and material creations that, together, inhabit the present.

This transformation of contemporary life has naturally affected archaeology, and directly so, in essentially two ways: in its study of the physical world around us and of humankind—through the remains of the humans who preceded us—and of the way things come down to us. We are now caught in a "crisis in European civilization" whose magnitude was revealed after World War II and which is manifest in the way the status of history has been undermined. As Hannah Arendt noted, with the irreversible "erosion of tradition," we no longer know how to regard a reality that has been so profoundly transformed by the cataclysms of the 20th century.[3]

The End of a Way of Perceiving Life

How is it that we can no longer identify with the aspirations of the generations who preceded World War I? Why do these people seem to belong to a time that has completely disappeared? Is it simply that their culture aged, and that a brighter, faster, purer one—ours—has taken its place? For philosopher of history Walter Benjamin, it goes far deeper than that: The 20th century witnessed the loss of our grasp of reality. In the new world generated by the massive industrialization of Western societies, the notion of human experience lost its meaning. In an essay entitled "The Storyteller," Benjamin forcefully explained how man's usual state of being had become incapable of dealing with the new reality introduced by mechanized warfare.

The "men who returned from the battlefield had grown silent—not richer but poorer in communicable experience . . . . For never has experience been more thoroughly belied than strategic experience was belied by tactical warfare, economic experience by inflation, bodily experience by mechanical warfare, moral experience by those in power. A generation that had gone to school on horse-drawn streetcars now stood under the open sky in a landscape where nothing remained unchanged but the clouds, in a force field of destructive torrents and explosions, the tiny, fragile, human body."[4]

Walter Benjamin was hardly the only person to make this observation. Because the unprecedented horror and suffering of trench warfare, where death blows were delivered on an industrial scale by literally "a-human" machinery, went so far beyond anything that had been experienced until then, the narratives and the very words used to recount experiences were unable to express what the survivors had gone through. And the result was that they did not express them. To be sure, the 1920s saw the publication of a significant number of first-hand narratives and war stories, but their veracity was immediately challenged. It was not that their depiction of events was exaggerated or magnified, but rather that the classical templates for rendering war experiences (with their fabled, literary, and ideological inflation: heroism, daring, patriotism, etc.) completely stripped of its meaning what had actually happened.[5] All the depictions *failed to capture* what the soldiers had lived through. The experience of industrialized warfare could not be told.

The systematic use of manufactured weapons and its corollary, "weapons of mass destruction," did not actually make war any more horrible than it had been until then, but the magnitude of its impact raised it to a level totally disproportionate to that of any previous war, thrusting the combatants into a "new reality" heretofore unknown, that of an industrialized world. After the initial shock of World War I, the collective traumas of the 20th century proved exceptional in that it was basically impossible to relate what had happened. In *If This Is a Man*, Primo Levi disclosed the existence, in the Nazi death camps, of certain "living dead," certain "mummies" who were referred to, in the jargon there, as "Muslims."[6] These were the people who had lost the will to live, or more precisely put, all hope. They knew they were going to die, and that there was no way to avoid it. For Levi, these people had seen "Gorgon Medusa." As the Italian philosopher Giorgio Agamben noted, the "Muslim" was the "quintessence" of the camps, the person who had seen it all but who, because he had seen what cannot be seen without immediately being struck dead, could not come back to tell about it. He was, in this respect, a person "whose humanity had been completely destroyed" and at the same time the ultimate witness to the Nazi hell that, in the minds of those who created it, no one was supposed to talk about.[7] There were no words to designate either the victims of the camps—in Nazi-speak, they were, at best, *Stücken*, "pieces"—or the act of killing them (which was referred to as "special treatment" and designated by the letters SB, for *Sonderbehandlung*). As for the people who carried out this "special" program, they had no more of an existence than did the "pieces" that they "treated." They were their equals, similarly "non-human," just fodder to be processed industrially by more or less automated machinery.

In Hiroshima, no one is left to bear witness to what happened near the point of impact of the atomic bomb. Absolutely nothing remains of the bodies, which were instantaneously vaporized. The people who were irradiated were known as *hibakusha*, literally "victims of the bombing." The term refers to the post-war wretches reduced to begging, whose stories were collected by the Japanese writer Kenzaburo Ōe. They too chose silence. The people of Hiroshima, one of them said, "prefer to remain silent as they await death. They want to have their own life and death . . . . I detest those who fail to appreciate how we feel about silence. We cannot commemorate August 6th; we can only let it pass away with the dead."[8]

The cataclysm that gave us Hiroshima raised to another level what Walter Benjamin called the loss of our grasp of experience, that is, the end of an age in which the physical world was the source of everything we knew. The "world since" is a contaminated place where the aftereffects of the cataclysm are still being felt. It has relegated the old world, the "world before," to so distant a past that it has become unreal, truly imaginary. In the world that followed the atomic bomb explosion, survivors were not people who had managed to escape and would soon resume "normal" lives; they were people pushed over the edge, outside the realm of the living. "I approached [the wall] because some low, dark figures seemed to be sitting in a row at its base," a young woman from Hiroshima said. "I could hardly tell man from woman or child, nor could I distinguish their ages. They were sitting in a line, practically naked, and all their faces and bodies had swelled up and turned brown, as if by common consent. One had already gone blind. Then I noticed a baby on someone's lap; the skin of its back was hanging down, like rotten, reddish loquat peeled off all around. Instinctively, I turned away from the sickening sight. They were all motionless, locked in a strange silence. It was hard to tell if they were alive or dead."[9]

On the night of July 28, 1943, Allied strategic bombing razed the city of Hamburg, creating a gigantic fire with flames rising over 6,000 feet into the air.[10] There were no words to describe what happened there; words didn't even have meaning in this context. In the Reich then in ruins, seven and a half million homeless wandered aimlessly from town to town. On trains packed with the fleeing refugees, some of the women carried the bodies of their dead children in suitcases. In his *Diary of a Man in Despair*, the German writer Friedrich Reck Malleczewen tells of a cardboard suitcase of one of the Hamburg refugees falling to the ground and spilling its contents onto the platform. "Toys, a manicure kit, singed underwear and last of all the roasted corpse of a child, shrunk like a mummy, which its half-deranged mother had been carrying about with her, the relic of a past that was still intact a few days ago."[11] As in Hiroshima, the devastating blows of the frightening world our world had become had stripped humanity of its human credibility:

"This wasn't the world; this wasn't humanity," someone who had seen the Warsaw Ghetto said. "I was not a part of it. I never saw such things . . . . I was told that these were human beings—*they didn't look like human beings.* Then we left."[12]

CHAPTER 5

The End of the World Has Already Come

In more recent times, the Russian journalist Svetlana Alexievitch collected the stories of survivors of the nuclear catastrophe that occurred in Chernobyl, in the Ukraine, on April 26, 1986. These survivors, too, who would subsequently die from radiation, found it impossible to relate their experience, for what had happened lay beyond all human experience: "No one knows what I have come back from," one of the "liquidators" of Chernobyl said. "And it's impossible for me to explain." Another person was clearer still. "We don't know how to make meaning of this experience. We just can't. *We can't place it in our experience as humans, or our human time frame.*"[13] And yet, the accounts of those who were able to talk about it evoked an extraordinary reality of new and fascinating beauty. Old Ukrainian peasant women referring to cesium spoke of "its jet black color . . . Pieces of it were lying on the ground, glowing." Children recalled "pools that were yellow and green, as if somebody had poured paint into them. We said it was pollen."[14] Curiously, these are the very colors that fascinated the Nazi camp survivors the first time they saw them in Treblinka, after the Nazis had begun cremating corpses on an industrial scale.[15]

Witnesses to the conflagration in Hamburg were also fascinated by the profusion of never before seen colors. "I watched the play of colors, fascinated," one of them said, "the red and yellow of the flames mingling against the background of the dark night sky and then separating them again. Even later, I never saw such a clean, brilliant yellow, such a bright red, such a radiant orange . . . I never again saw such deep, bright colors, not in any painter's work. And if I had been a painter myself . . . it would have taken me a lifetime to find those pure colors."[16]

All of these people were struck by the transformation of the physical world they had witnessed. In speaking of the way the world was "turned upside down" during World War II, Paul Virilio, an urbanist, said that what came to him on the morning he woke to see the city he was born in bombed to the ground and open to the sky "wasn't so much the horror of the people buried alive in cellars . . . as the city's sudden transparency, the plainly visible changes in the urban environment, and the motility of things inanimate and immobile."[17] Suddenly, the setting in which reality unfolded had become completely, dizzyingly, different. It had entered the realm of the fantastic, with displaced, strangely familiar objects present in a physical world that had been totally transformed. The most basic points of reference had lost their everyday meaning—or at least had found another—while a normally invisible reality was now everywhere apparent. This was new. In Hiroshima, thousands of swallows with seared wings hopped painfully along the ground. In Chernobyl, there were dead sparrows "lying around everywhere. They'd be raked up and taken away in containers, along with the leaves."[18] In Hamburg, in the days following the conflagration, there was an unbelievable proliferation of "huge and iridescent green flies such as had never been seen before. They swarmed in great clusters on the roads, settled in heaps to copulate on ruined walls, and basked, weary and satiated, on the splinters of windowpanes. When they could no longer fly they crawled after us through the tiniest of cracks, and their buzzing and whirring was the first thing we heard on waking," the German writer Hans Erick Nossack said.[19]

Fundamentally, there are no words available to us to relate whatever it is that these new worlds represent in our minds. After *Shoah* came out, it became commonplace to say, as Primo Levi had, that the experience of the camps could not be told. The same can be said of the mass destruction of German cities carried out by the Allies in 1944.[20] "There were true stories," told by survivors, noted German writer Wilfried Gerhard Sebald, "that exceeded anyone's ability to grasp them." They were essentially incommunicable, for, in unprecedented fashion, they told of "a world that could no longer be presented in comprehensible terms."[21] "An event has occurred for which there is neither a way to describe it, nor an analogy nor a previous experience" to which it can be related, Svetlana Alexievitch said about the catastrophe in Chernobyl. "Neither our eyes nor our ears nor our vocabulary are adequate . . . . In this particular instance, *our previous experience is clearly insufficient.* Ever since Chernobyl, we have been living in a different world, and the old world no longer exists."[22]

The old world, the world of human beings, is outdated. In the world since, it has become obsolete, as has "our entire culture," which, as one of the Chernobyl survivors said, "is just an old trunk with a bunch of stale manuscripts."[23] All that remains are ruins and debris. In Chernobyl, one of the psychologists sent to help the survivors said, "my past no longer protects me."[24] The past doesn't mean anything anymore. It no longer holds the present. In this present that is ours, the only things that remain from the past are ruins and vestiges, debris that is barely distinguishable from *trash*. Our time, the time of history, is henceforth *here and now*, in other words, the present, which is basically where archaeology functions.

Aside from the shared impossibility of relating experience, the catastrophes of the contemporary era have unsettled our grasp of reality. Here again, this phenomenon is not in itself new (it is the very nature of catastrophes), but, having occurred on an industrial scale, it has opened onto a new reality that has become the reality of the industrialized world. The world that was transformed by industrialization can no longer be expressed in the same terms; its stability, and thereby its veracity, are no longer assured. In the wink of an eye, our physical world can be thrown into the ineffable. We live in a "society at risk," whose advent coincided, according to the German sociologist Ulrich Beck, with the shift from an industrial to a postindustrial age that began around the middle of the 20th century.[25] As Beck noted, the threats hanging over us involve "hazards and insecurities that have been induced and introduced by modernization itself; . . . in contrast to all earlier epochs (including industrial society) the risk society is characterized essentially by a lack: the impossibility of an external attribution of hazards."[26] Thus, the worst, when it occurs, is produced and managed industrially, which makes it fundamentally unpredictable, so that, for "societies at risk," the possibility for catastrophe is essentially defined in terms of statistical probability. What is has been dissolved into what might be, into what should not normally occur, but what might nonetheless happen at any time, and this affects all of physical reality: With the coming of the postindustrial age, the way we interpret what exists has been thrown into crisis.

What Is Happening (To Us)?

We have not forgotten the peremptory judgment made by the German philosopher Theodor Adorno, in the aftermath of World War II: "Any cultural creation that comes after Auschwitz, including the urgently needed criticism of Auschwitz, is just a pile of crap." The import of the catastrophes of the 20th century is so radically "anti-human" that it is impossible to situate them in the normal flow of our common history. Like a needle on a scratched record, history is stuck in a groove. And because it has failed to move on, it has revealed the vast field of ruins that is our world today, and our identity as survivors standing amid the disparate, decomposing remains of a past that has become fundamentally incomprehensible. The "loss of [our grasp of] experience" that Walter Benjamin spoke of is the consequence of our radically changed relation to the physical world that was brought on by industrialization and automation. We are now cut off from our past by what W. G. Sebald very rightly characterized as a "deficiency in transmitting history" which, paradoxically, exposes us to a repetition of dehumanizing catastrophes whose causes may always be different, but whose consequences remain essentially the same. There will be other Hiroshimas, other Hamburgs, other Chernobyls, and, inevitably, other Auschwitzes too, be it here or elsewhere. The process of repetition, the fact that experiences can no longer be inscribed in history, has radically altered what time as experience represents. Entire chapters of our collective history have been erased. In saying that "my past no longer protects me,"[27] the psychologist at Chernobyl was speaking for us all. Nothing we transmit will have a place over time for, as we know, experiences in which mankind is stripped of its humanity will continue to occur—they have never stopped occurring—in a transparent time with neither past nor future, that is the *real* time of the postindustrial age.

In his schizophrenic madness, Friedrich Nietzsche had clear insight into what loomed on the horizon as the shift toward industrialized war, which is to say total war, got under way in Europe during the final decades of the 19th century. "The treasures in books have become obsolete; thus, an eternal recurrence is possible *today*," he wrote in the summer of 1882.[28] Nietzsche had already perceived very clearly that "the treasures in books," the priceless knowledge common to all mankind, was now just a heap of vestiges—what a survivor of Chernobyl called "a trunk of stale manuscripts"—whose contents were totally "obsolete" in that this knowledge had been outstripped by the unprecedented, incommunicable reality of our own world. If we have effectively reached this point, if knowledge of the past can no longer be transmitted to help us grasp reality, then the world is entirely within the here and now, in a "nowness" that has no place in history, which is to say neither memory nor the possibility of evolving. For Nietzsche, this nowness, that is perpetually begun anew, was the distinctive mark of the "aimless time" in which we live.[29] "The world persists," he added. "Nothing is becoming or happening. Or better still, something is becoming, is happening, but it has never begun becoming and will never cease happening . . . . It is *preserved* by the two processes."[30] (emphasis added)

We need to pause and reflect a bit on Nietzsche's remarks on "recurring time," whose concept he was among the first to formulate with the phrase "eternal recurrence," and which Walter Benjamin's "nowness" (*Jetztzeit*) echoes. A few terms need defining. As Georgio Agamben noted, the Nietzschean idea of the "eternal recurrence of the same" (*ewige Wiederkehr des Gleichen*) has to be understood in its etymological sense, not as the identical reproduction of events that once took place in the past, but, on the contrary, the reiteration of *resemblances* between phenomena, whether they occur in the past or the present, or are still to occur in the future.[31] In fact, it is not just formal resemblance that is important here, but rather the process of reiteration (*Wiederholbarkeit*), which, for Nietzsche, was fundamental to the workings of the *eternal recurrence*, as an eternalizing phenomenon.[32] (emphasis added) In the ahistorical time of man's condition as Nietzsche saw it, what was fundamentally at work was simply a kind of *potential for repetition* that brought present, past, and future to coexist through their shared resemblance or, one could say, the similar forms they assumed. Herein lies the origin of the particular sense of time that leaves its mark on all things archaeological. As Gilles Deleuze so brilliantly explained:

> How can the present be constituted in time? How can the present pass? The passing moment could never pass if it were not already past and yet to come—at the same time as being present. If the present did not pass of its own accord, if it had to wait for a new present in order to become past, the past in general would never be constituted in time, and this particular present would not pass. We cannot wait, the moment must be simultaneously present and past, present and yet to come, in order for it to pass . . . . The present must coexist with itself as past and yet to come. The synthetic relation of the moment to itself as present, past, and future grounds its relation to other moments. The eternal recurrence is thus an answer to the eternal problem of *passage*. And in this sense it must not be interpreted as the return of something that is, that is "one" or the "same." We misinterpret the expression "eternal recurrence" if we understand it as "return of the same." It is not being that returns but rather the returning itself that constitutes being insofar as it is affirmed of becoming and of that which passes. It is not some one thing which

returns but rather returning itself is the one thing which is affirmed of diversity or multiplicity.[33]

We need to grasp something very profound about time and history that Deleuze has brought to light here. As the art historian Georges Didi-Huberman noted, since it is the present that passes, the past is shaped *within the present*.[34] It is clear that in our physical world, material objects exist only as "aging things," so to speak, or, to use Walter Benjamin's well-known image, as an "accumulation of ruins." And this explains why the materiality of our present is made of the past. It follows that, as opposed to the classical perception of historicists, the past does not stand in opposition to a present that would be situated elsewhere, ahead of the past. The past is rather the very stuff of the present, at the same time that it holds, in the present, the seeds of what is to come, as surviving traces of what has been. This is what defines the things man makes, and gives specific historical meaning to what we recognize as the typological attributes of material creations. Once we become aware of this, the world, as it appears to us and in its materiality, takes on new meaning, and its history reappears as potential memory.[35] No longer can we perceive history as a succession of chronological periods or distinct time frames. We see it now as a dynamic unfolding over time that is radically different from the time of conventional history, which has effectively become "obsolete." It becomes the subject of an as yet unnamed discipline that it remains for us to create, largely from the abandoned matter lying in the field of ruins of the present.[36] This unnamed discipline is none other than archaeology.

Once one accepts that the workings of time as we now understand them do not adhere to a principle of sequence, but rather to a principle of reiteration, the very nature of events—as triggering factors or steps in particular processes—comes under challenge because traditional, "obsolete" time (the time of a pure, linear, chronological sequence of events) is necessarily continuous, whereas a present time of reiteration is filled with gaps and necessarily discontinuous. In fact, it is above all made of gaps, absences, and missing elements. Paradoxically, this is a sense of time that we archaeologists are particularly familiar with. Every occupied site, even the most continuously occupied, is a series of hugely discrete archaeological events separated from each other by more or less lengthy periods of time in which nothing happened, at least nothing we would call archaeological "phenomena." Nietzsche's writings help us grasp why, fundamentally, the discontinuousness of time is consubstantial with the notion of events; that is, with what he defines as acts in time. "It is *impossible* for an act to take place between two consecutive moments," he said, "for then these two points in time would coincide. Every act is *actio in distans*, there is always a leap." And he added: "As for how an act *in distans* is possible, I have no idea."[37] Nor do we. We know nothing about it, but we recognize that the repetition of archaeological phenomena in a given place has meaning as an archaeological construct developed over time only because events are discontinuous. There is even the sense that it is their discontinuity, the gap between events, that shapes the identity of archaeological "phenomena" as such. For example, with regard to some network of ditches that periodically has to be redrawn and redug—because they tend to be wiped out and disappear—their recreation progressively allows each part of the ditch to acquire over time a particular morphological or structural identity, on the basis of which it becomes possible to construct a history or an evolutionary pattern. The time line (Nietzsche uses the term *Zeitlinie*) is thus far from being regular. It is shattered into a multitude of points, comprised of unconnected events whose various temporal dynamics (Nietzsche speaks of their *action*) are inextricably superimposed: "Time is not a continuum," he wrote in the caption to his chart of temporal dynamics. "*There are just points in time; there is no line. Actio in distans.*"[38]

Time in Crisis, History in Crisis

Thus, the crisis in history that confronts us is symptomatic of a crisis in time that the cataclysms of the 20th century, such as those involving mass destruction, have brought to the surface of the present. The present is a field of ruins, where past and future lie inextricably entwined, dismembered, and crushed in the present. The crisis has affected our overall representation of reality. All of 19th- and 20th-century Western culture derived from an 18th-century concept that placed nature, society, and individuals within an historical framework. "Locating things in history," which was fundamental to the social sciences that were developed in the 19th century, was the product of a reversal in perspective.[39] Up until that point, the dominant form of history related tales of dynasties, the purpose of which was to provide a mythical basis for royal authority and the social order based on nobility under the *Ancien Régime*. In other words, an order that had been created in the past had to be constantly recalled in the present, lest it fade away, which it had a tendency to do; the core of historical time lay in the past. The 18th-century concept of history stood in diametrical opposition to this post-medieval approach: Past events (such as the Frankish conquest, which was the basis for nobility) were still resonating in the present, fueling social and political struggles, which were essentially historical in nature. The import of history was no longer found in the past—in *origins*—but in the present, *now*.

The new conceptualization gave rise to an array of disciplines characterized by their study of phenomena as they evolved *over time*, through the various transformations they had undergone, rather than in the act of their creation. These disciplines were fundamentally *historical*. Aside from history itself—said to be the preeminent discipline of the 19th century—there were archaeology and prehistory, geology and paleontology, philology, and even the history of economics, to cite just the principal ones. Each in its own way, they focused on areas that had been opened by "locating things in history." Nonetheless, as Michel Foucault noted in *Les Mots et les choses*, the rise of these disciplines coincided with a break in traditional historical time that had linked the history of mankind to the evolution of the world from as far back as earliest Antiquity.[40] In effect, in the first half of the 19th century, advances in geology and paleontology brought to light the fact that nature had its own history and that, above all, it was wholly independent of the history of mankind, if only because it predated it. In the second half of the 19th century, discoveries in prehistory and the emergence of an archeology that dealt with civilizations that preceded those of classical Antiquity also derived from the new sense of historical time. Not only did these new fields lie beyond what was then known of human history, they functioned independently of it. Researchers had not simply discovered the existence of a history "predating history," be it in the evolution of species, the transformation of the products of man's industry, or the development of languages, they had concluded, more importantly, that each of them followed its own evolutionary pattern. Their evolutionary chronology was, as Michel Foucault said, "a function of their individual nature"[41] and not of the "single" time of classical history. Historical time had exploded into a multitude of chronological dynamics that could not be synthesized.

Thus, at the end of the period that saw the development of new fields of study that had come with the 19th-century creation of new historical disciplines, man actually found himself increasingly estranged from his history. As a historical phenomenon, the history of mankind now appeared to be just one of several. Not only did these other histories not necessarily share a common meaning with that of mankind, but also, and above all, they were not subordinate to it. With the systematic attempt to "locate things in history" that marked 19th-century research, the specter of a radical "de-historicization" of existence loomed even larger: Man was no longer the lone and necessary source of the events he experienced. As Michel Foucault noted, man, as the subject of historical study, found himself fundamentally "exposed to the event,"[42] naked and defenseless in the face of a history that evolved beyond his control and was self-sustaining itself. No longer did he create history with his technology, his civilizations, and his empires. He had unwillingly become the product of his past, haunted still by his most distant and most obscure origins, which is what Freudian psychoanalysis would demonstrate compellingly in the first half of the 20th century. But if man were no longer the master of history's destiny, what was? What were the forces at work? One of the essential undertakings faced by the disciplines developed in the 19th century was that of re-establishing the links in historical time. What was needed was a common thread that could reconnect the various elements of the history of the world, which had become a fragmented mass of individual histories.

But where was one to find that thread? In their search for coherence, 19th-century researchers looked to man himself, to the living, working, thinking being. They reasoned that it was because man consumed, produced, and used what he found in his environment that these numerous histories had emerged, for however different they were from each other, they were all nonetheless connected to him. They reasoned further that it was because man thought and reacted that he could give direction to these transformations; in other words, that he could subordinate them to human progress. As the history of mankind unfolded, human society grew more complex, and tools became increasingly efficient; eventually, nature would fall entirely under man's control. Prehistorians found evidence for these ideas in the redefined chronology of man-made objects that clearly showed uninterrupted progress through technological advances. To their minds, these were facts that could be as clearly substantiated as the scientific principles on which they rested. A new tale relating the history of mankind could at last be substituted for the one relating the history of kings. This was an objective history, based on historical sources, one that would replace an imaginary one rooted in myth. Nineteenth-century researchers had only to get down to work.

An Aborted Attempt at Normalizing Historical Time: Prehistorical Archaeology

Nowhere among the new historical disciplines that were developed in the 19th century do we find a greater concentration of the hopes and disappointments associated with the search for the basic unity of the history of mankind than in the archaeology of prehistoric times. It was a new approach to the study of man's origins that combined the disciplines of geology, anthropology, and archaeology in an attempt to join the history of Man to the history of Nature. As was the case for the other disciplines that emerged in the 19th century, archaeology and the study of prehistory drew on Enlightenment concepts that were at once scientific and ideological:

1. *The past is fundamentally knowable.* Archaeology has posited that it is possible to establish the chain of past eras because their sequence obeys an internal logic, that of the uninterrupted development of human society. This is the postulate Gabriel de Mortillet formulated at the end of the 1860s as *The General Law of the Progress of Mankind*[43];
2. *The past is fundamentally understandable.* Archaeology has posited that it is possible to understand social behavior and the reasons why given techniques were adopted by "primitive" or "prehistorical" man because, as men, we share with them a common identity. This was Gabriel de Mortillet's *Law Known as the Psychological Unity of Mankind*;
3. *The past can be fundamentally represented as it was.* Archaeology has posited that it is possible to depict past eras because each of them has its own identity, which is effectively a function of their place in history. This was Gabriel de Mortillet's *Law of*

> *Similar Development,* which held that, in the course of their historical development, all human systems pass through analogous stages.

At the same time that the traditional notion of a "single" time of history was exploding into a multitude of notions of historical time, prehistorians were pursuing their search for a synthetic overview that could bring together under one common concept of historical time—namely, that of human progress—the cultural time of the history of civilizations and the natural time of nature's evolution. The new notion of prehistorical archaeological time was to unite the time of the social sciences and that of the natural sciences. It was to be both a specific, that is a single, notion of the historical time of civilizations and societies throughout the evolution of mankind, and a notion of time for the natural world governed by the scientific laws of the human species but one that functioned in ways similar to the "real" laws of nature. This search for a combined form that had been envisioned in the 1870s never came to fruition, for the two notions were incompatible: The "cultural" time of the development of civilizations was firmly rooted in a series of discrete sequences whose identities depended on their place in history, whereas the "natural" time of nature had no specific place in time; it resided entirely in "nowness." There was no way to introduce this new notion of natural time into the old notion of cultural time, at least not without stripping it of what defined it, by reducing Darwinism, for example, to a simple matter of transformation. As a consequence, and as was true for all the historical disciplines that developed in the 19th century, archaeology and the study of prehistory continued to be based on a traditional concept of historical time marked by the following characteristics:

1. *Time is unidirectional.* History advances from the past toward the future, from its origins toward its rise and its culmination, from simple to complex. Any movement from complex to simple or from conclusion to (re)commencement is by definition anti-historical;
2. *Time is linear.* History is guided by the principle of the unity of place and action. A multiplicity of events can be reduced to a

series of essential episodes that are inscribed within an overall dynamic that is effectively the movement of history itself;

3. *Time is causality.* Linear time is the basis for a history that, as it unfolds, develops its own internal coherence. In this perspective, historical time—which overlaps with history being carried to a conclusion—has direction. It naturally leads to the completion of processes.

The historical disciplines of the 19th century found themselves in much the same situation as the "exact" sciences, with the growing realization that time and space might not function in conventional ways: The multiplication of "historical subjects" had created the impression that it might be possible to relate all aspects of reality by systematically linking all of them together, like netting. But there was one hole that stubbornly refused to close, and history slipped through it, carrying off, through what had been at the outset a minuscule opening, larger and larger sections of conventional historical time. In the social sciences, the problem with the net of historical time appeared in the form of psychoanalysis; in the natural sciences, that of Darwin's theory of evolution. In fact, the more one tried to normalize and unify these new notions of historical time, the less well things went. It is revealing that these attempts failed most decisively in the new fields of study opened by prehistory and archaeology:

1. Finds from prehistoric times gave proof that the past was not entirely knowable, since no one had suspected the existence of some of the things that were discovered, in particular human fossils (but could they really be said to be human?) whose disappearance dated back to immemorial times;

2. The past was not really understandable, since it was extremely difficult to explain the presence of some of the artifacts unearthed. The problem lay in the classical historical sources that were relied on to associate artifacts with events and practices they had nothing to do with, in particular because they belonged to far more remote eras;

3. As a result, the past could not really be represented because, in basing representations on history, artifacts were placed in the wrong time frames. The most flagrant example was that of archaeologists working for the "National Antiquities" under Napoléon III, who attributed pre-Roman artifacts to the time of the Gallic Wars, when what had actually been unearthed were entire segments of the material culture of proto-historic times.

Thus, in spite of—or more accurately perhaps, because of—the great increase in historical knowledge in the second half of the 19th century, the past proved to be, to use the British art historian David Lowenthal's expression, a foreign country.[44] Discoveries made by the new historical disciplines, in particular archaeology and prehistory, revealed that the histories through which we pass are not actually entirely ours, which is to say that they are not exclusively about us. Other forces, which lie beyond our control, also have an impact on them. As Bergson so insightfully demonstrated at the turn of the 20th century, the recording of past events is shaped not so much by the history to which they bear witness as by the matter in which their memory is inscribed.[45] For both the social and the natural sciences, separating memory from history was a challenge to the traditional notion of historical time. But these revelations—that came out of the new disciplines of memory, foremost among them, prehistory and archaeology—were co-opted, so that they failed to undermine the view that held that man was the lone agent of history. It would fall to the industrial cataclysms of the 20th century to revolutionize the notion of historical time.

History Is a Void

So, we must look to the 19th century for the origins of the crisis in history generated by the "loss of experience" wrought by the mass destructions of the 20th. At the very same time that history was diversifying into a series of specialized disciplines—including prehistory and archaeology—it was becoming incapable of relating history; that is, of giving an account of what happens over time. In his theses on "The Concept of History," Walter Benjamin perceptively identified the characteristics of this crisis, which essentially corresponds to representing the present as "what is happening":

> The tradition of the oppressed teaches us that the "state of emergency" in which we live is not the exception but the rule. We must attain to a conception of history that accords with this insight. . . . The current amazement that the things we are experiencing are "still" possible in the 20th century is *not* philosophical. This amazement is not the beginning of knowledge—unless it is the knowledge that the view of history which gives rise to it is untenable.[46]

Benjamin stated it clearly: The point in criticizing conventional history is to redefine it in a way that will free us from the oppression in which the repeated, massive disasters of our age have held us. These disasters are not anomalies in the "normal" course of history. They are rather, as Benjamin noted, the rule. At issue is the entire idea of viewing history as progress, not only because it licenses these disasters to happen, but above all because it sanctions them. We need to understand the workings of historical time in a different way, for given the repeated breakdown in mankind's humanity, the situation is, indeed, untenable. In truth, we have no choice. If we wish to hold onto that part of humanity that the industrialization of our lives is taking from us, we have to free ourselves from the historicist's approach to history.

What does "history viewed as progress" actually mean? It is not just a matter of how we interpret the movement of history; it involves, at a deeper level, the way we connect past events through cause-and-effect relationships. And that is because, in the traditional perspective, progress "will be made when the time comes," because the sought-after ideal is constantly being pushed further back into the future, and because it is constantly altered as progress marches on. In this view, progress follows the gradual, regular movement of time, so much so in fact that it has become, in the operative archaeological chronologies dating back to the 19th century, the very measure of time. But, as Benjamin emphasized, linear time seen as progress is hollow. This historical time is nothing more than the industrial time of clocks, reading everywhere alike at any given time, perfectly homogeneous. "The concept of mankind's historical progress cannot be sundered from the concept of its progression through a homogeneous, empty time," he said.[47]

Viewed as progress, the linear time of history is merely the chronological basis for ordering events in an unbroken series that purports to relate history. As a result, the reconstruction of the past undertaken by the various historical disciplines reduces to an exercise in accumulating facts, in filling in the missing events and periods which, when juxtaposed, will eventually constitute the thread of universal history: One single history of man's progress, similar everywhere in the world, across one universal time. It would be hard not to recognize here the traditional approach to archaeology. Ever since the beginning of the 18th century, the archaeology we have come to know has been collecting artifacts which, when stuck into what Walter Benjamin referred to as "empty, homogenous" time, provide us with matter for the history of mankind and the history of civilizations. The work we do in archaeology has never been anything else. It is fundamentally historicist in that its sole aim is that of reconstructing a universal history of the progress of human society from its most remote prehistorical origins up to the flowering of the civilizations of recorded history. But this conventional approach is inconsistent. It neither has nor can it yield a theoretical basis for a history that would allow us to understand, or to imagine, what transpires over time. As Walter Benjamin said: "Historicism rightly culminates in universal history . . . . Universal history has no theoretical armature. Its procedure is additive; it musters a mass of data to fill the homogenous, empty time."[48]

We know this well: an archaeologist's work consists of filling in with new artifacts whatever "holes in time" remain on the line of archaeological periods that has been so painstakingly drafted over the course of what is now more than a hundred and fifty years. For we are completely at a loss when spaces are devoid of sites or artifacts; we don't know what to say. We don't know what is missing at certain points, and the gaps, which break the continuity of known facts, make it impossible for us to fit together the unconnected pieces we do have. Our job is effectively one of drawing pictures to give an identity to moments we deem discrete.

First among them are typo-chronological pictures that allow us to identify, for each defined sequence in time, a corpus of objects produced at the same time in different places. Next are pictures of proto-historical spaces with countless farms and elevated fortifications, followed by those of Roman spaces, the perfectly aligned urban centers, the quadrangular centuriations in the countryside, etc. These are all just so many pictures—by definition, homogeneous—lined up according to their place in time, or rather proceeding from each other in a purely linear fashion. Here again, this is an approach peculiar to the historicist's view of the past.

"Historicism," Walter Benjamin wrote, "contents itself with establishing a causal nexus among various moments in history. But no state of affairs having causal significance is for that very reason historical. It became historical posthumously, as it were, through events that may be separated from it by thousands of years. The historian who proceeds from this consideration ceases to tell the sequence of events like the beads of a rosary. He grasps the constellation into which his own era has entered, along with a very specific earlier one. Thus, he establishes a conception of the present as nowness shot through with splinters of messianic time."[49]

This passage is especially important for archaeology, for it underscores an essential property of archaeological matter: memory. The long periods of time over which artifacts take on meaning undermine the postulate on which the traditional, historicist approach to the discipline is based. We know well that just because some archaeological phenomenon—for example, the construction of a series of houses or the grouping together of several tombs—took place at some time, it does not follow that it gave rise to a series of other archaeological phenomena. Archaeology offers numerous examples of abandoned sites whose abandonment could not have been predicted from the archaeological phenomena themselves. Rather, this grouping of tombs or that cluster of houses was simply the last to occupy those sites, which, consequently, were no longer used for the same purpose. Above all, as Walter Benjamin noted, we know that archaeological phenomena can come back into play; that is, can become, at a much later date, the cause of other archaeological constructs. The linear trajectory posited for transformations over time comes tumbling down, incapable of accounting for the true nature of archaeological evolution. Once this has been grasped, the chain linking archaeological phenomena ceases to be essential; what is essential is rather the interaction between phenomena across differing spans of time and space: the way in which given constructs help shape subsequent ones and, inversely, the way in which subsequent constructs represent retranscriptions of those that preceded them. Left behind is the realm of history, where what matters is to know what "really" happened. We have entered the realm of memory. What matters is the impact past phenomena have had.

In effect, "the historian who subscribes to this view ceases counting off historical phenomena like beads on a rosary." He has realized that the place of past events is not simply the moment when they occurred. The entire approach to history is upended by this change in perspective, as is especially evident in archaeology. Whereas the conventional historicist approach works through *addition*, it is the nature of the archaeological approach to work through *construction*. Archaeological constructs are not necessary links on a vast linear chain; they are the nexus of an immense variety of possible connections that it has become critical for us to explore. As Walter Benjamin said, they offer "a revolutionary chance . . . to blast a specific era out of the homogenous course of history."[50] In effect, once artifacts are recognized to be "material memory," they are liberated from the "homogeneous course of history" in which the traditional, historicist approach would channel them; linear history made of a series of events loses all meaning. Once limited to the laborious reconstruction of a chain of events, the task of the historian, and more specifically of the archaeologist, has become that of seeking out connections between events over periods of time.

"Nowness"

If an archaeological phenomenon's place in time is not simply the moment when it happened, where is it? The answer to that question is unsettling, contrary to common sense. Its place is everywhere and nowhere, or if you prefer, archaeological constructs are unconcerned with time, even if they are fundamentally a product of time. Actually, their place is the present, understood as ongoing time, "nowness." Here again, we have to free ourselves from the linear or historicist perspective of historical time and focus on the memory generated by the interaction between archaeological constructs over time. As we have already seen, every archaeological construct, be it an object or a structure, has to find its place. In this respect, it is the product of the give and take between old and new, between what exists and what is to come. It is the interplay of this configuration that gives archaeological constructs their particular identity, and because they are constantly being altered (through aging or disappearance, renovation or change), the identity of archaeological constructs is unstable; in other words, they evolve. The give and take happens at whatever particular moment it occurs, because time always functions as "nowness."

To carry this further, note that archaeological constructs are reiterated (structures have to be repaired, remodeled, or replaced) because "nowness" is constantly begun anew; it is always the same and never the same. As we have already seen, our own relation to artifacts plays out in the present. Not only are artifacts discovered in our present, but, above all, the meaning we attribute to them is shaped by their relation to our present. For example, the mass graves from 20th-century conflicts are fundamentally the same as those of more distant periods, but the answers we look to find in them are a function of our relation, here and now, with the past events with which they are associated. And, a week from now, we may find that they have a different meaning, not so much because we will interpret them differently as because new events will have connected them to historical processes that we are not yet aware of. This is why, today, in the mass graves in the former Yugoslavia, we are working to identify the individuals buried there, whereas in those of the Neolithic period we rather seek the reasons behind the massacres. Once again, it is in "nowness"—our time, today, and for the people who will come after us, tomorrow—that archaeological remains take on meaning as memory. As Walter Benjamin noted, "history is the subject of a *construction* located not in homogeneous, empty time, but in *time filled full with "nowness."*"[51]

Thus, at every moment in time, remains from the past function in the present as memory. This argument, which holds that not only are physical fragments of the past a part of the materiality of the present but that they continue to have an impact in "nowness," opens the way to a fundamental re-evaluation of the notion of history. It provides the means with which to understand differently the evolutionary processes at work in historical time. As opposed to what is implicit in the historicist's approach to time, the future is not constructed in a linear manner with innovations, if only because the present is "filled" with the past, whether consciously or not, whether it has been recognized to be the past or not. The future is part of an ongoing process; it is shaped by the present. As archaeology clearly shows, it is because artifacts are always present that they can endure and evolve over time. Thus, if the present is filled with the past, the inverse is equally true. The present is not only that which is happening right now, it is what has always been going on: the aging of matter, the erosion of sites, the imprints bodies make on places—in short, the effects of constantly renewed life as it takes shape in the present, right before our eyes, as it did in all the presents that preceded us and will in those that follow. In effect, far from bearing the stamp of uninterrupted change, the present rather bears that of an "eternal return to itself," always like itself in its diversity as in its singularity. As Walter Benjamin said, all of time is effectively "filled full with nowness." Clearly, we are dealing with a different approach to time: "a unique experience with the past," as Benjamin put it,[52] that is ours to grasp in every moment of time whose artifacts we discover. "Articulating the past historically," he said, "does not mean recognizing it 'the way it really was.' It means appropriating a memory as it flashes up in a moment of danger."[53]

Opening Time

Walter Benjamin's principal idea is that the present, as "nowness" (he uses the term *JetztZeit*) can at any point establish a connection with some moment of the past. In so far as the evolutionary process is also a return to the past, either as the past is reactivated, or as what exists is re-evaluated, the present moment holds the potential for evolution in all its forms. This is how we have to understand what Walter Benjamin meant in saying that there exist in the present "pieces of messianic time," meaning time that allows for anything to happen, right then, "filled" time that holds memory, as opposed to the "hollow" time of the historicist.

Walter Benjamin sought to liberate historical time, to open the little box in which historicism had locked it away. For if the future is open to innumerable possibilities because it is production, then the past is as well, because it is memory. Every present moment opens simultaneously onto a multiplicity of possible futures and a variety of probable histories. They are fundamentally constructions. Given this, it is wrong to believe that there is necessarily only one history, for we don't know how the archaeological constructs we opt for and produce will evolve, or the time spans over which their evolution will play out. Walter Benjamin effectively was urging us to reject traditional, historicist schemas and undertake a (re) reading of history that goes, as he put it, "against the grain."[54] And this is a highly unsettling approach, for it challenges our understanding of history. Our starting point, or basis, is no longer the past, but the present, the very place in which we find ourselves, and in which time finds itself too. This is an exasperating approach for historians and more generally for all those who define themselves as authorities on the past, because it inverts our relation to time and, in doing so, upends what we took to be the order, literally, of things.

Benjamin's understanding of history, as he formulated it in reaction to the crisis in modern civilization, has a host of direct implications for archaeology because, like the approach to archaeology, his approach to history draws on the materiality of vestiges that bear witness to the past. His radical thinking leaves us with the following alternative:

- We can believe that archaeology essentially functions as a kind of para-history, re-establishing typo-chronological sequences that it orders in time so as to reproduce periods of culture and the evolutionary processes of civilizations. In this case, archaeology would deal with the past alone and have nothing to do with the present, which by its nature would be essentially non-archaeological. This is the traditional approach, the one the discipline accredits, as it were, and precisely the one that Walter Benjamin denounces with the term "historicism";
- Or, we can believe that archaeology deals with the material memory of the past and that it is the work of the archaeologist to study the way in which memory is constituted over time, in which case the present, understood as "nowness," would become the locus for interpreting the past. This is precisely the approach promoted by Walter Benjamin, as a way out of the impasse into which historicism has led.

Formulated in ethical terms, Benjamin's questioning of the approach to history speaks to all disciplines that deal with the past, for it addresses the role and the definition of *archival matter*, which is the intermediary through which the past becomes known to us. Let us not forget that it was by considering what was held in their archives from the point of view of the present that geologists and paleontologists were able to open the locked historical time that had fettered natural history, and that continues to fetter archaeology. The change in perspective is upsetting because it calls into question concepts that are basic to our conventional—"modern," Bruno Latour would say—understanding of the world and the people around us.[55] This is especially so because one of the defining characteristics of modern Western tradition is situating nature and men in history. What is being challenged is not so much the processes of historical transformation as the interpretation of the data on whose basis history can be (re)constructed. In effect, the archaeological artifacts that constitute the archives of the past definitively lose their status as witnesses and become traces, signs. On a deeper level, the cataclysms of the contemporary era, by turning the present into a field of ruins, have revealed the nature of these traces to us. It is in the residue, the trash, the detritus that we must seek the archival matter that truly bears the imprint of the vanished past.

Notes

1. Francis Fukuyama, *The End of History and the Last Man* (New York: Free Press, 1992).
2. Eric Hobsbawm, *L'Âge des extrêmes. Histoire du court XXe siècle* (Paris: Complexe, 1994).
3. Hannah Arendt, *Between Past and Future: Eight Exercises in Political Thought* (New York: Viking, 1972).
4. Benjamin, "The Storyteller," in *Selected Writings.*
5. Jean Norton Cru, *Témoins. Essai d'analyse et de critiques des souvenirs de combattants édités en français, de 1915 à 1928* (Paris: Les Étincelles, 1929); Jean Norton Cru, *Du témoignage* (Paris: Jean-Jacques Pauvert, 1967).
6. Levi, Primo, *If This Is a Man* (New York: Collier Books, 1961).
7. Giorgio Agamben, *Remnants of Auschwitz,* trans. Daniel Heller-Roazen (New York: Zone Books, 1999), p. 133.
8. Quoted in Kenzaburö Öe, *Hiroshima Notes*, trans. David L. Swain and Toshi Yonezawa (New York: Grove Press, 1996), pp. 19–20.
9. Quoted in Öe, *Hiroshima*, pp. 176–7.
10. The Hamburg survivors refused to speak. Janet Flanner, writing for *The New Yorker* described them as "a dwindled population in black and with bundles—the silent German people appropriate to the city," quoted in W. G. Sebald, *On the Natural History of Destruction*, trans. Anthea Bell (New York: Random House, 2003), p. 31.
11. Quoted in Sebald, *Destruction*, p. 29.

12. Claude Lanzmann, *Shoah, An Oral History of the Holocaust.* (New York: Pantheon, 1985), p. 174.

13. Quoted in Svetlana Alexievitch, *Voices from Chernobyl*, trans. Keith Gessen (London: Dalkey Archive Press, 2005), pp. 82, 90.

14. Quoted in Alexievitch, *Voices*, pp. 28, 221.

15. Richard Glazar, a Treblinka survivor said: "In a flash, the whole countryside, the whole camp, seemed ablaze . . . . And from the window we kept on watching the fantastic backdrop of flames of every imaginable color: red, yellow, green, purple . . . . This was the first time it happened. We knew that night that the dead would no longer be buried, they'd be burned." Lanzmann, *Shoah*, p. 14.

16. Quoted in Sebald, *Destruction*, p. 86.

17. Paul Virilio, *L'Insécurité du territoire* (Paris: Galilée, 1993), p. 16.

18. Alexievitch, *Voices,* p. 222.

19. Quoted in Sebald, *Destruction*, p. 35.

20. In so far as the old distinction between *friend* and *foe* loses its meaning, given the magnitude of the disaster, Gert Ledig's *Payback*, which relates the experience of the Allied bombing of a German town in July 1944, is also worth reading (London: Granta Books, 2003).

21. Sebald, *Destruction*, pp. 23, 10.

22. Svetlana Alexievitch, *Tchernobylskaïa molita* (Moscow: Ostojié, 1997), p. 33.

23. Alexievitch, *Chernobyl*, p. 123.

24. Alexievitch, *Chernobyl,* p. 27.

25. Ulrich Beck, *Risk Society: Towards a New Modernity,* trans. Mark Ritter (London: Sage Publications, 1992). The evolution of modern societies has also been noted by the British sociologist Anthony Giddens, *Consequences of Modernity* (Cambridge: Polity Press, 1990).

26. Beck, *Risk Society*, pp. 21, 183.

27. Alexievitch, *Voices,* p. 27.

28. Friedrich Nietzsche, *Fragments posthumes sur l'éternel retour,* trans. Lionel Duvoy (Paris: Allia, 2003), p. 46.

29. Nietzsche, *Fragments,* p. 52 (notes from November 1882 to February 1883).

30. Nietzsche, *Fragments,* p. 93 (notes from spring 1888 for "the new vision of the world").

31. As Agamben said, "In the eternal recurrence, there is something like an image, a resemblance . . . something like a total image or, in Benjamin's terms, a dialectical image. Only thus can the eternal recurrence assume its true dimensions." Giorgio Agamben, *Image et mémoire. Écrits sur l'image, la danse et le cinéma* (Paris: Desclée de Brouwer 2004), p. 98.

32. In a rough draft of the outline for a work on the "Return of the Same," dating from 1881, Nietzsche marked down as one argument for the part including the explanation of the phenomenon of "eternal recurrence": "a wholly different eternalisation . . . the eternal core, eternal reiteration, has to be included in it" (Nietzsche, *Fragments,* p. 40).

33. Gilles Deleuze, *Nietzsche and Philosophy,* trans. Hugh Tomlinson (New York: Columbia University Press, 1983), p. 48.

34. Georges Didi-Huberman, *L'Image survivante. Histoire de l'art et temps des fantômes selon Aby Warburg.* (Paris: minuit, 2002), p. 171.

35. As Nietzsche said in *Fragments posthumes sur l'éternel retour* "once we have arrived at this thought, colors all change, there is a new *history*" (Nietzsche, *Fragments,* p. 40).

36. The conclusion to *History, The Last Things before the Last* (New York: Oxford University Press, 1969, p. 8), Siegfried Kracauer's unfinished work on history, was to be "the message [that] is hidden away in the interstices" between conflicting dogma and thereby "a way of thinking and living." Kracauer emphasized that the approach he had taken throughout his work essentially involved "the rehabilitation of objectives and modes of being *which still lack a name and hence are overlooked or misjudged,*" p. 4.

37. Friedrich Nietzsche, fragment XXVI, 12 in *Œuvres philosophiques complètes,* trans. P. Rusch (Paris: Gallimard, 1990), p. 317.

38. Nietzsche, *Œuvres,* fragment XXVI, 12, p. 318.

39. Michel Foucault analyzed this change in depth in lectures given at the Collège de France in 1976. *Cours au Collège de France 1976–1977* (Paris: Gallimard, 2004).

40. Michel Foucault, *The Order of Things. An Archaeology of the Human Sciences,* trans. Taristock/Routledge (New York: Pantheon Books, 1970), pp. 367–73.

41. Foucault, *The Order,* p. 368.

42. Foucault, *The Order,* p. 370.

43. Gabriel de Mortillet, "Essai d'une classification des cavernes et des stations sous abri fondée sur les produits de l'industrie humaine," in the journal *Matériaux pour l'Histoire primitive et naturelle de l'Homme,* 1869, no. 5, pp. 172–9; Gabriel de Mortillet, *Le Préhistorique. Antiquité de l'Homme* (Paris: Reinwald, 1883).

44. David Lowenthal, *The Past Is a Foreign Country* (Cambridge: Cambridge University Press, 1985).

45. Henri Bergson, *Matières et mémoire* (Paris: Presses universitaires de France, 1896).

46. Benjamin, *Selected Writings,* vol. 4, p. 392. Siegfried Kracauer spelled out the basis for Benjamin's criticism of the universal time of historicists: "The idea

of a progress in humanity is untenable," he said, "mainly for the reason that it is insolubly bound up with the idea of chronological time as the matrix of a meaningful process." Kracauer, *History*, p. 150.

47. Benjamin, *Selected Writings*, vol. 4, pp. 394–5.

48. Benjamin, *Selected Writings*, vol. 4, p. 396.

49. Benjamin, *Selected Writings*, vol. 4, p. 397.

50. Benjamin, *Selected Writings*, vol. 4, p. 396.

51. Benjamin, *Selected Writings*, vol. 4, p. 395.

52. Benjamin, *Selected Writings*, vol. 4, p. 396.

53. Benjamin, *Selected Writings*, vol. 4, p. 391.

54. Benjamin, *Selected Writings*, vol. 4, p. 391.

55. Bruno Latour, *Nous n'avons jamais été modernes. Essai d'anthropologie symétrique.* (Paris: La Découverte, 1989).

CHAPTER 6

RAGMEN OF THE PAST

Dear Discoverer...

The past has something to tell us; it would like us to hear its story:

> Look everywhere, under every clod of earth. Buried in the ground are dozens of pages that I wrote along with others so that people will know what happened here. There are a lot of teeth, too. We were the ones, the *Kommando* workers, who intentionally buried them all over this place, as much as we could, so that the world might have tangible proof of the millions of people that have been killed here. As for ourselves, we have given up all hope of seeing the Liberation.[1]

This is one of the hand-written notes that have been discovered in Auschwitz since 1945. Pieces of paper were folded or rolled up inside tin cans or bottles and either buried outside the crematoria or stuck in with the piles of human remains. They are known as the "Auschwitz Scrolls."[2] The last of them was unearthed in the 1980s. Others surely exist, although a good many of them were probably destroyed in the immediate aftermath of the war by Polish peasants rooting about in search of rumored "Jewish treasure." These writings are already filled with the holes wrought by time; eventually, they will all disintegrate. One of them reads:

> We will continue to do what we have to. We are going to [blank] everything and hide [blank] the world but simply bury in the ground and in [blank]. But any person who wants to find, still [blank], can still find

> [blank] of the yard, not towards the street [blank] on the other side, you'll find a lot over there [blank] because we have to, as we have until now, until [blank] event [blank] keep letting the world know everything as in a historical chronicle. Starting now, we are going to bury everything in the ground.[3]

At Auschwitz, the *Sonderkommando* was supposed to keep the wheels of the extermination machinery turning by burning the dead bodies and getting rid of their remains. They themselves would ultimately fall victim to this machinery. They knew well that to bear witness to "what happened here," they had to bury things in the ground, to hide them in the hopes that one day "good" men would find them and hear what they had to say. "Good" men capable of recognizing the truth and making it known: that is what the victims of Auschwitz hoped that we would be. They placed in us the only hope they had, that of making their "story" known to the world, but they understood that they had to take us by the hand, for we would really not know "what happened here." They would have to tell us exactly where to look and exactly what we had to find: through their writings, of course, that chronicled their story "like a history book," but, above all, with material evidence of the extermination. Because we would not know what that evidence consisted of, or how it was to be read, they had to tell us that buried around the grounds were piles of teeth that they had taken from as many of the bodies they could get to unseen, so that we could see that these were human teeth, taken from the masses of people who had been exterminated here.

One member of the *Sonderkommandos*, Alter Szmul Fajnzylberg, wrote that he had hidden "a camera and a metal can with gas in it" near the crematorium in Birkenau.[4] The camera—which has not been found—probably held a roll of film with pictures of the extermination process. The can undoubtedly contained what had been used to gas victims before their remains were burned in the crematoriums. Here, too, we had to be told. Without the writings of the men of the *Sonderkommando*, we would not have known that there were pictures of considerable historical importance in that fossil of a camera, or that that rusted can actually contained what remained of the gas used at Auschwitz. We could not know, and the victims of Auschwitz were counting on our goodness and our perseverance. There was nothing else they could do but hope that we would take care of the vestiges they left for us—even though they had no material value—and that we would tirelessly "look everywhere, under every clod of earth," even though we would not know exactly what it was we could expect to find. They hoped that we would be attentive to every bit of debris from their past, all of which had meaning for them. They inscribed this prayer on a tattered piece of paper: "We ask of destiny *Yehi rotsn milifneikho, eyno shoymea kol bekhies*: do us at least this favor—*shetashim dimeosseinu benodeikho lihies*: hide these pages of tears in the gourd of life, that they may fall into good hands and reach their *tikoun*, their destination."[5]

A few years prior to this, in Thesis II of "On the Concept of History," Walter Benjamin had written that: "[T]here is a secret agreement between past generations and the present one. For our coming was expected on earth. Like every generation that preceded us, we have been endowed with a *weak* messianic power, a power on which the past has a claim. Such a claim cannot be settled cheaply. The historical materialist is aware of this."[6]

But our preparation for such responsibility is inadequate. We do what we can, but we don't see much. We do not possess the keys to the past, which is foreign to us. We don't know what we are supposed to see. To our eyes, there is just a pile of rubble: a jumble of bits of brick, flagstone and white porcelain, pieces of concrete and rusted metal that we cut our hands on. Proust, too, was convinced that dead souls were locked inside things or beings, and that they spoke to us; that they were there for us. In the episode of the three trees, as Proust was looking out the window of the carriage carrying him across the countryside, he sensed that their silhouettes were familiar to him:

> I chose rather to believe that they were phantoms of the past, dear companions of my childhood, vanished friends who were invoking our common memories. Like ghosts they seemed to be appealing to me to take them with me, to bring them back to life. In their simple and passionate gesticulation I could discern the helpless anguish of a beloved person who has lost the power of speech, and feels that he will never be able to say to us what he wishes to say and we can never guess . . . . I watched the trees gradually recede, waving their despairing arms, seeming to say to me: "What you fail to learn from us to-day, you will never know. If you allow us to drop back into the hollow of this road from which we

sought to raise ourselves up to you, a whole part of yourself which we were bringing to you will vanish forever into thin air."[7]

Anyone who has ever worked on a dig knows the feeling very well.

What Do Archaeologists Look For?

What does archaeology involve, and where do its roots lie? To date, the history of archaeology has essentially been written by archaeologists themselves, which is why this "history of archaeology as seen by archaeologists" tends to overlap with a historiography of research. Technical advances, the fields worked in, the approaches, in a word, the discipline's present scientific identity is what defines the framework for composing its history, both as a historical process (the development of archaeological research) and for given historical moments (such and such a period of the "origins of archaeology"). In other words, the present situation of archaeology represents its culmination. By working backward, one can establish archaeology's progress from its very first baby steps in the past to its supposed apotheosis in the present.

What we are challenging here is the very nature of the historiography of archaeology, along with a certain conception of time and historical evolution whose impact on the way in which we deal with artifacts has already been noted. As the historian and philosopher Michel de Certeau pointed out, "historiography is in effect a narrative that serves . . . to make a point."[8] We need to pause and reflect on the particular light this sentence casts on the specific kind of historical writing that historiography represents. Historiography places past events in a narrative context. The narrative supplies the internal logic, the supposedly self-evident sequence of historical facts that gives coherence to the historical demonstration supposedly produced by the historical exercise. It is thus, effectively, no more than argumentation. Historiography does violence, as it were, to historical facts in order to support the point of view that the author, as scholar—or the community of scholars to which he belongs—espouses with regard to his discipline. As Michel de Certeau said, historiography is built on linear time, with a series of sequences "leading towards the intended recipient's, that is, the reader's present time."[9] With its linear perspective, which we know to be flawed, historiography sanctions not only the discipline's present situation, but more specifically those groups who dominate the discipline and who are (or who aspire to become) the field's authorities. This is why appeals to the history of archaeology are the practice of researchers seeking historical legitimacy for their perception of the discipline.

This approach says nothing about the actual subject matter of a *history* of the discipline; that is, the effect time and durations have had on the way methods and practices have evolved. For that, we have to look not to the linear, sequential time of historicists, but to the "time filled full with nowness" that is peculiar to archaeological thinking. More specifically, we need to bring to light another strategy for historical time, to use Michel de Certeau's phrase.[10] As he has shown, historiographical discourse is based on a division that is, in truth, arbitrary, one that separates present from past[11] by positing the present as the place where research is carried out, as its starting point, so to speak, while past events are the subject of historiographical studies and the matter with which to piece history back together "more or less as it happened." We know that this is not how things work. The simple need to write a "true" or "objective" history is ample proof that the past continues to haunt the present, just as we know that the ubiquitous manifestations of that need for truth—those thousands of archaeological publications, digs, and collections—continue to inhabit present time. Here, as out in the field, the present is filled with vestiges of the past, and the interplay of their durations is far more important than the actual "events." This is a radical approach. In focusing on duration, we open a breach in the wall that has been erected between us and the past. In effect, we abandon traditional historiography and root it in another approach that requires us to listen to the distorted echo of the past that lies in our present.

Events on one side, durations on the other: We are effectively dealing, as Michel de Certeau pointed out, with two opposing ways of "apportioning the space of memory"; in other words, of conceiving the relation between past and present. We need to push this distinction as far as possible. We have here two antithetical views of time and history: disciplines (psychoanalysis and, by definition, archaeology) in which time plays an integral role and that recognize the presence of the past in the present, and other disciplines that are based on events (historiography, as well as traditional history and archaeology) and that posit the coexistence of past and present as objectively impossible. Further still, for traditional historical disciplines, the validity of historical elucidations is based on an objective ordering of events that define each other mutually. For disciplines dealing in durations, on the other hand, history is built on replication, with various sites or events reiterated, regenerating new ones in more or less identical or altered forms. Similarly, in the conventional approach to history with its linear, sequential time, the relation between events is correlative, with events acting only upon those that are closest to them, whereas for disciplines that deal in durations, sites, or events can combine with or replace each other. Acknowledging that the interplay between durations is the stuff of archaeology involves more than simply patching up the current history of archaeology; it undermines the very explicative structure of conventional history and, in particular, the notions of effects and disjuncture (two events separated in time cannot be the same thing) that are fundamental to the historiographical perspective. On the contrary, for an approach that deals in durations, certain events can act in place of others, while specific sites or events that are normally imperceptible can suddenly reappear front stage in some new identity. In recognizing the unwavering persistence of durations, we slowly undo the knot of this fundamental ambiguity of history.

How does one recognize and interpret these durations working at the deepest levels of the archaeological discipline? One has first to dismiss all thought of grasping how they function simply by sequencing the evolution of the discipline as historiographers would. The tension between durations becomes apparent at those moments when knowledge is being passed on, when the definition of the discipline is both reaffirmed and transformed. The way archaeology is taught, or the debates that ensue when it is suddenly confronted with unexpected discoveries, says a lot more about the true nature of the discipline than one might imagine. Some new discovery comes along and challenges the prevailing system of representation; it demands an immediate response, either rejection or assimilation—which one matters little. What does matter is the revealing way the discipline—archaeologists—reacts in the face of this challenge. What is accepted and perhaps even given prominence, and what is rejected because it is judged unacceptable or illegitimate? And when it is not a matter of some external element challenging the nature of the discipline, what are the practices by which it is reaffirmed in the classroom, that other realm in which we define archaeology?

The Search for Pictures

As opposed to other disciplines based on the observation and analysis of ground holding a fossilized memory of the past (such as geology, paleontology, and pedology), archaeology is not taught through the fieldwork that is done where the material recording of the past lies buried. The dig—which is essential, as André Leroi-Gourhan pointed out, in so far as archaeological investigation into the past destroys the object of its study as it proceeds—is not part of the archaeology curriculum taught in universities, which have never subscribed to it. Archaeology is rather taught as the history of civilizations, or the study of the stylistic traits of material cultures of the past. In effect, from the 18th century on, the discipline has basically involved expounding upon the cultural creations of ancient civilizations. The close ties between art history and archaeology, which most archaeologists today fail to see the need for, date back to the earliest times.

For archaeologists have never been excavators. They trade in elaboration, in piecing together and analyzing images of the past. The definition of archaeology lies in this practice of accumulating depictions of the material remains of the past from as far back, in fact, as the discipline's most distant origins, in Antiquity. Traditionally, an archaeologist's principal occupation has never been digging down into the earth or exhuming artifacts from lost civilizations. It has rather been that of examining catalogue illustrations, building plans, cross-sections, and profiles. An archaeologist is a drawing-room savant, an iconographer searching for morphological details that are characteristic of past civilizations that he then compares. He is a solitary scholar living in a world filled with lists and inventories, a man adding to the "paper museum" conceived by the great antiquarians of the 18th century whose aim it was to inventory, illustrate, and describe every artifact or, to use their phrasing, every monument to the industry and arts of past civilizations. The flurry of archaeological activity in the 19th century grew directly out of this encyclopedic archaeology of the Age of Enlightenment. Far too often, we picture the scholars of the period from 1860 to 1900 as men toiling with primitive means on random digs, whereas in truth their work principally involved putting together a gigantic iconographic corpus of artifacts from all over the world. The *archaeological albums* from the second half of the 19th century, with their magnificent watercolors, were designed to enhance the illustrated catalogue of mankind's past material creations. To do so, scholars did not hesitate to make use of the most advanced techniques of their times, in particular photography, which was still in its infancy. Twentieth-century archaeology has maintained the tradition, enriching it further through computer science, from the end of the 1980s on. And everything would seem to suggest that 21st-century archaeology will continue contributing to this "complete archaeological collection," especially with the documentary resources now available through digital imaging.

Antiquarians of the classical age initiated a tradition of renderings of—or, more precisely, the rendering of—artifacts. From the outset, in the second half of the 16th century, it was through depictions of sites and remains that the first archaeological discoveries became recognizable as such. Highly realistic drawings and watercolors revealed the shapes and features of vestiges from long-gone eras. In the very first representations of digs, done by Augst at the end of the 16th century, we see ruins of a Roman theater that are still recognizable today, with its typical arena seating and rubble masonry. Similarly, in Lucas de Heere's watercolor from 1575, we immediately recognize the circular ditches and round of crumbling megaliths at Stonehenge.[12] The realistic depictions of artifacts that began appearing, like foreign bodies, in contemporary landscapes went hand in hand with the newly acquired awareness of the existence of different eras of the past: Since man-made objects from ancient times were clearly different from those of their own era, their distinctive features had to be included if they were to be recognized as such. Shortly thereafter, they would realize that the past itself included different eras that also had to be distinguished, each with its own morphological characteristics. That would be the agenda for archaeology in the second half of the 19th century and throughout the 20th.

Thus, a basic concept was introduced with the very first depictions of archaeological matter: The past is fundamentally different from the present, with an identity of its own. As we have already noted, this is one of the basic assumptions of historicism, and it is clearly related to the late 16th century's awareness that vestiges of the past were "other." In this perspective, the proliferation of "ultra-realistic" renderings of artifacts in the 19th century (objects shown not only in great detail, but with the effects of aging visible in colors and materials that had been subject to corrosion) was hardly the work of chance. Such depictions had become the foundation for the essentially historicist perception of the history of mankind.

Proof of this can be found easily enough by looking at archaeological renderings in the medieval tradition, where depictions function in a wholly different manner, totally absorbed in the nowness of the present. One recurrent image is that of vases that have spontaneously arisen from the ground. One encounters them as late as the end of the 15th century, in Barthélemy de Glanville's magnificent *Livre des propriétés des choses* (*Book on the Properties of Things*). Alongside various animals in lairs burrowed into the soil or far below the surface of the water, we see vases breaking through the earth's crust, popping up like mushrooms.[13] The vases, which resembled those in daily use in the 15th century, actually represent vases from ancient times, which could be either the Bronze Age or Roman times; there is no way to tell. As opposed to depictions from the end of the 16h century on, we are nowhere in particular. The late-medieval rendering of vases emerging from the ground is metaphoric. The picture is saying: Just as certain animals come out at night from the ground in which they sleep, so objects emerge wholly formed from the earth. For the past had not yet been recognized as an entity of its own, fundamentally apart from the present.

This is a highly resilient view of time. It reappears in the 18th century, in particular to explain how fossils are formed. For example, a great number of scholars believed that seashell shapes and fish remains embedded in rock could be attributed to a particular property, a *vis plastica* or *virus lapidifica*, that produced, in stone, calcified animals whose species were similar to those in the living world.[14] Even Voltaire wondered "if the earth wasn't capable of producing these fossils" that looked like sea animals, but were found far from the sea, embedded in mainland rock.[15] Clearly, both for animals in stone and pots rising from the earth, this "fabulous" explanation for the presence of objects from the distant past was linked to a truncated representation of time. In the "flat time" that preceded the "Deep Down Time" that 19th-century geologists gave us, the past was entirely contained in the here and now, and whatever objects bore witness to it were lodged in the present like supranatural entities that could be explained only as the result of extraordinary happenings. Until archaeological vestiges attested to the existence of other periods, megaliths could only be understood as "giants' tables," prehistoric funeral mounds as the "tombs of fairies," and Roman ruins as "Sleeping Beauty castles." Recognizing a time frame specific to the past was a first step, one that the advent of archaeology would give concrete expression to. It was only at the beginning of the 19th century, with the revelation of the massive sweep of ancient times, which was essentially geology's doing, that other explanations for these mysterious phenomena gained credence. The extraordinary nature of these anomalous remains—which were by then seen as vestiges of the past—was attributed to the cumulative effect of gradual phenomena that played out over very long stretches of time and eventually resulted in radical transformations. It is easy to understand why it has been so difficult to criticize the historicist tradition from within since it is so deeply rooted in the conventional approach to archaeology. It has come down to us from far back in the past, from the very source from which archaeology emerged as a discipline.

Ragmen

The appearance on the scene of archaeologists, and of historians as well, coincided with the recognition that the memory of the past held a particular meaning that was not immediately accessible and thus not immediately understandable; reading it required elucidation. From as far back as the Roman era, the archaeologist has been that solitary man obsessed with a past that has been lost to us. We find him in vacant lots, dumps, and cemeteries, looking for what is hidden, exhuming what lies buried in the ground, gathering together what had been scattered, and, like a ragman of the past, stirring up piles of dead things in search of debris from mankind's ancient past. His field is wherever the detritus of what once lived has been discarded. The subject of his study, which he shares with no one, is the remains, or more precisely, the rejects of history. The archaeologist has always been an *antiquarius*, an antiquarian, a collector.

The first archaeologists from classical times were conscious of their role as rummagers in the ruins of the past, in the original sense of the term. It fell to them to identify the remains on which they could build their knowledge of ancient times, which is to say to define the raw material of archaeology. At the beginning of the 18th century, the Benedictine monk Bernard de Montfaucon proposed the clearest of definitions: "I have reduced," he said in his *Antiquity Explained, and Represented in Pictures*, "into one Body all Antiquity. By the word Antiquity I mean only that which may be the Object of Sight and may be represented by Figures."[16] The raw material of archaeology is everything of the past that can be seen; today, we would say everything whose existence is physically recorded. Thus, to speak of archaeological data is to speak of visualizations: in the 18th century, drawings and maps; beginning with the 19th, photographs; today, digital imaging based on the physical properties of matter. All of this, Montfaucon said, was to form a corpus (today we would say a field peculiar to archaeology). The path to follow was clear: Archaeologists were to study observable artifacts and record them through various renderings. It was an agenda that, from as far back as the 18th century, we have not strayed from in the slightest.

And yet, we know that something is missing, that in itself this approach will not suffice. The problem is the pictures. No picture, however true to the original, can ever speak by itself. As opposed to what Montfaucon suggested, amassing the largest possible collection of representations of artifacts is not enough to give us a complete vision of the past. As he himself pointed out, even if we try to reduce our vision to what we can see (drawing plans of buildings, pictures of divinities, types of tools and attire, etc.), it won't be enough. Pictures always require elucidation to tell us, at the very least, what is being shown. It was surely not by chance that it was an expert in the visual arts, Caylus, an art critic trained in painting analysis, who developed the catalogue concept that went hand in hand with that of an iconographic corpus. As a man with experience in drawing and engraving, and as a member of the Royal Academy of Painting and Sculpture, where great art collectors called upon his expertise, Caylus was well versed in representational art.[17] His masterwork, the seven volumes of the *Recueil d'antiquités égyptiennes, grecques, étrusques et romaines*, published between 1752 and 1767, is an immense annotated catalogue of the pieces from Antiquity that could be found in France in the middle of the 18th century.[18] The method Caylus used for his inventory drew on the principles used for cataloguing art works. The present-day approach to describing archaeological objects is merely a continuation of his.

CHAPTER 6

They Didn't See a Thing

Archaeologists first ventured into the field in the first half of the 19th century. Their initial steps out of their studies were particularly half-hearted. It wasn't at all clear to them that digging down into the ground was actually necessary. Up until the 1860s, when the dig became normal practice for collecting archaeological matter, archaeologists had been content to take advantage of building sites where ancient objects of interest to them had been unearthed. When they did go out into the field, they did so to pick them up at the source, before the workers who had brought them to the surface could damage or discard them. At the outset, then, the point of fieldwork was essentially to collect objects. And consequently, collectors or purveyors had to be hired, which immediately opened the way to forgeries and trafficking, as Boucher de Perthes, in all his naiveté, would sadly find out in Abbeville.[19] In fact, until at least as recently as the 1950s, when stratigraphic digs became common practice, archaeologists did not have the means with which to verify the authenticity of archaeological finds. The extraordinary turn taken by the "Glozel affair" stands as proof that, as late as the 1930s, the discipline lacked the tools with which to determine whether there actually existed an authentic archaeological context for the "Paleolithic writing" found in the "Field of the Dead" near Glozel, in central France.[20] Not until archaeologists became capable of "reading" the ground did they gain significant command of it.

Behind every archaeologist stands an antiquarian. In his eyes, a dig is just another way of collecting works from times gone by; in this case, at the site at which they lie buried. So we dig, but what exactly for? For a very long time, archaeologists had absolutely no idea. And they had no idea because they didn't know how to read terrain. They lacked the knowledge that could tell them where to dig to find the things that interested them. They were led to certain places either by someone's chance discovery or by the inferences they drew from historical works. The extraordinary growth in archaeology during the "National Antiquities" period of the 1860s that prompted research on sites of the Gallic Wars is actually most notable for the thorough misunderstanding of archaeological terrain that it reveals. As early as the 17th century, topographical details of the itinerary followed by the Roman legions as related in the *Commentaries on the Gallic Wars* had led people to believe that the defeat of the Helvetians had taken place in the Côtes de Beaune region, possibly near the village of Cussy-la-Colonne, in Burgundy, where a column from the Roman period, decorated with narrative scenes, might, it was thought, commemorate the event. Archaeologists immediately interpreted the great number of funereal mounds located there as "Celtic" tombs that had been raised to bury the thousands of slain Helvetians.[21]

So, digs were begun on the "Chaumes d'Auvenay" plateau in Ivry-en-Montagne, first in the 1840s, then at the beginning of the 1860s. Everything archaeologists laid their eyes on seemed to support their conjectures. Because certain finds, in particular bronze pins, were consonant with Bronze Age lacustrine sites recently discovered in Switzerland, they confidently concluded that the people buried on the Auvenay plateau were Caesar's Helvetians. Similarly, they attributed the seemingly random organization of the funerary mounds, and the fact that burial vaults were occasionally placed on top of each other, to the haste with which the survivors had had to bury their dead. A similar scenario played out in Alise-Sainte-Reine, also in the department of Côte d'Or. The discovery of a store of bronze weapons and tools in 1860[22]—which was immediately likened to the Auvenay finds—triggered methodical investigations that eventually led researchers to the trenches Caesar had dug around Alésia. Chance had it that the trenches ran just a few yards from the hoard of bronze objects found on the Épineuse farm, which is how they came to be discovered. The interest in Alise-Sainte-Reine essentially stemmed from a scholarly tradition, also dating back to at least the 17th century, that placed Caesar's Alésia there. Nobody realized that the funerary mounds in Ivry-en-Montagne and the bronze weapons in Alise-Sainte-Reine predated the events ascribed to them by almost a millennium. And people essentially failed to see that because exhumed objects were considered no more than material illustrations of a history for which written works were a better source.

The major obstacle hindering the development of digs was the inability of archaeologists to detect anything whatsoever in the ground other than the objects they had been trained to look for. Thus, when Claude Rossignol, who was to become the head curator of the Museum of National Antiquities, was led out to the Auvenay plateau in Ivry-en-Montagne in

May 1842, he at first failed to notice anything. He had brought with him a dozen workers recruited from among the peasants in nearby villages. In the course of just one day, they opened ten prehistoric funerary mounds, but Rossignol didn't see a thing: "Everything has been tossed together," he wrote, "bones, earth, pebbles and ashes. There are human remains at every level, but it's a horrible mess throughout."[23] Twenty years later, the site stirred the interest of archaeologists in Napoléon III's entourage, who were convinced that the tombs were those of the Helvetians routed by Caesar. Félix de Saulcy went back to the dig site, but the initial work done in 1859 didn't reveal anything of interest to him either. Beginning in 1860, the presence of well-preserved skeletons allowed him to identify burial chambers, but he never identified those in which they had been cremated, which we now know were present in far greater numbers.[24]

The early digs of the 19th century produced mixed results everywhere, not because they were not well organized, but because archaeologists were unable to discern any order whatsoever in what they found. This is what happened, for example, in Sainte-Colombe-sur-Seine, in Burgundy, in 1845–1846, when the Maîtres, wealthy landowners from Châtillon-sur-Seine, had the monumental "La Garenne" tumulus leveled. Half of the mound had already been removed when the laborers suddenly uncovered a great number of bronze and iron objects. The head librarian in Châtillon, Dr. Bourrée, rushed out to study the pieces, and above all to sketch them. Today, we know that this exceptional find was part of an opulent tomb containing a Celtic wagon dating from the 6th century BCE—the first one of these found in Europe—and that it almost certainly held other things as well, in particular the remains of one or more bodies. Nothing of that appears in Dr. Bourrée's account. He too mentions the haphazard organization and the jumble of remains being disinterred: "Beneath the debris (of a cauldron)," he wrote, "there lay a huge muddle of iron rods—some straight, some curved, some whole, some fragmented—broad nail-heads, three legs with ends shaped like claws for some kind of instrument, the remains of an iron hoop partially covered with thick strips of engraved bronze, three of which taper into handles."[25] Today, we know that these corroded, crushed fragments were part of the remains of a large Etruscan tripod that supported a bronze cauldron with griffin-headed protomes that probably stood next to a large four-wheeled wagon covered with iron plates.

In short, in Sainte-Colombe, as in Ivry-en-Montagne, no one saw anything. They couldn't identify anything, and they didn't understand anything. Twenty years later, once again in Sainte-Colombe, Eugène Stoffel, who coordinated the archaeological work on Gallic War sites for Napoléon III, directed the excavation of the nearby monumental La Butte mound. They were secretly hoping to find the tomb of one of the major Gaulish figures in Vercingétorix's coalition. In 1863, the dig unearthed a burial vault located on top of a four-wheeled wagon entirely covered with iron plates. Only then did it become possible to identify the bewildering metallic debris described by Dr. Bourrée as the remains of a similarly iron-plated wagon. For it is the recurrence of certain discoveries and similar observations that allows archaeologists to identify the often unrecognizable objects they find buried in the ground. And that can only happen over time. Archaeology is in truth uncharted territory that we have yet to fully explore.

Let Nothing Slip Away

Initially, archaeologists ventured into the field only periodically. They would generally show up at the end of the workday to see what had been unearthed and gather the pieces they wanted to take back to their studies. In truth, they were not especially interested in archaeological "context," which, for them, was just a vicinity. Objects were the focus of attention. And because objects were not constantly being dug up, they did not deem it necessary to be present while the workers were seemingly just moving the soil around. This was the prevalent situation, not only in France, but in all of Europe up until World War I, after which archaeologists began camping on-site to observe what was going on. It was not until these antique enthusiasts became capable of recognizing specific indications that were contained in the land, and that had to be gleaned in a particular manner lest they be lost, that they began actually participating in digs. In Germany and England, the practice would not become prevalent until the 1930s, and in France not until the 1960s. For a long time, archaeologists had recourse to a foreman, most often some local they could trust, someone who knew the land well and who would do the looking and the digging for them. Nonetheless, the principal concern of this new breed of archaeologist now out in the field remained the discovery of objects. They took spade in hand because they did not want to run the risk of seeing valuable objects, "of scientific importance," shattered or discarded.

Archaeologists are basically trained to identify archetypes and recognize classes of objects and monuments, not to glean information from the land in which artifacts are buried. Field archaeology has been a self-taught practice, just manual know-how, for as long as it has existed. It is, in fact, highly significant that for a very long time there wasn't a single work available that dealt specifically with how to proceed on an archaeological dig. In France, methods explicitly aimed at taking the terrain into account rather than simply extracting artifacts from the ground did not appear until 1950, with the writings of André Leroi-Gourhan.[26] In the Anglo-American world, excavation techniques didn't make their way into publication until the 1970s. Like the general overviews that began appearing in France in the 1980s,[27] most of these were essentially a compilation of techniques aimed at an audience of amateur archaeologists and beginners. They explained how exhumed artifacts were treated and the procedures used to analyze them; nothing was actually said either about the land in which these pieces had been fossilized or the way it needed to be approached. Explanations of an actual dig continued to go wanting.

And yet, writing field manuals was an old practice dating back to the end of the 19th century. The first of them appeared in 1890, for the most part in conjunction with research being carried out in eastern France.[28] All of them used the same approach: locating the specific places where artifacts, especially prehistoric artifacts, could be found and explaining how their shape could help identify the different types that belonged to the periods known to archaeology. Thus, archaeological sites like caves, burial mounds, and prehistoric "hillforts" were basically considered to be particular kinds of locations in which to find specific types of objects from certain periods of the past (in the German sense of *Fundstelle*). There was no discussion of the way they lay, their internal structure, or the different transformations these sites might have undergone over time. The purpose of the manuals was to help people find artifacts.

Thus, as the practice of digs developed, what archaeologists gained from the land essentially reduced to greater antiquarian knowledge. With the new artifacts raised from the ground, they further added to the "complete museum" that had been envisioned in the 18th century. But whereas 18th-century collections displayed only certain kinds of objects, which were generally selected for esthetic reasons, 19th-century digs multiplied artifact types and promoted the importance of both identifiable debris that could be pieced together and of remains, however uninteresting their appearance. It was also as a result of digs that there emerged the ill-defined notion of archaeological context, as it quickly became clear, shortly after the turn of the century, that certain kinds of objects were associated with certain structures, most notably certain types of tombs. Last, it became clear from increased fieldwork that certain morphological and stylistic traits had cultural and chronological significance; that the kinds of sites and structures where objects were found were linked to specific civilizations and ages. These new aspects of the corpus of artifacts had to be described, and above all depicted, in maps, drawings, and photographs. This was the knowledge that Joseph Déchelette attempted to compile in the second decade of the 20th century, in his *Manuel d'archéologie préhistorique et celtique*. Albert Grenier would extend the content to include the Roman period, in work he carried out until 1960.[29]

And yet, as we know, digging down into the land provides a wealth of new information, which had begun appearing in Europe in the 1860s, that has no place in archaeological collections or museums. Digs were providing quantities of previously unknown, generally unremarkable remains that were not works per se. They were fragments of vessels, bones, stones, pieces of wood, even unrecognizable metal debris. At first, archaeological museums—the case of the National Museum of Antiquities is exemplary—put entire collections on display, the component pieces having been selected by collectors or archaeologists right in the field. Ceramic shards and bones that were not deemed of interest were generally left where they were found. Then, faced with a growing mass of remains not worthy of display (up until around World War I, these were referred to generically as *rejects*), museums began setting aside storage space for them. Today, these remains represent the essential part of finds, but museums don't have the storage capacity to house them. Digs have effectively undermined the notion of museums and collections, which represented the form through which antiquarian knowledge was both generated and displayed.

CHAPTER 6

Knowledge of the Terrain

Reduced to the search for objects characteristic of a particular age, archaeology found itself at a dual impasse. It inundated museums with artifacts they had no space to store, and directly contributed to their decontextualization by overlooking whatever information found in the terrain that was not directly relevant. Observations on the particularities of archaeological terrain came rather from the study of the complex structure of archaeological sites.

In truth, it would be a very long time before these observations gained currency in established archaeological discourse. In the 1880s, the English archaeologist Pitt Rivers began methodically recording information about terrain with the idea of reconstructing, after the dig, in as much detail as possible, the original layout of uncovered archaeological structures.[30] To Pitt Rivers' mind, the ideal form of reconstruction was a model. It could recreate in three dimensions a filled-in network of trenches or leveled foundations, which maps and cross-sections could render only partially. It is highly significant that Pitt Rivers' work faded into oblivion during the entire first half of the 20th century. It was the British archaeologist Mortimer Wheeler who recognized his seminal role in the development of stratigraphic methods and helped establish the technique in the 1950s through work done on a prehistoric site in Maiden Castle, England.[31]

Stratigraphy is fundamental in recognizing signs of the archaeological memory recorded in the ground, but the approach was not the product of prehistoric archaeology, but rather of the classical archaeology practiced in the Mediterranean basin. In Greece, in particular, archaeologists found themselves confronted with the complex archaeological structures of buried cities from ancient times. One of the first Anglo-Saxon works to deal specifically with the methodology of an archaeological dig was published in Cambridge, in 1915, by an English classical archaeologist, P. J. Droop.[32] Every archaeologist, Droop said, "should know . . . that what he finds is not more important than the conditions in which he finds it."[33] In Droop's view, the sea-level altitude at which finds were made was of major importance in situating them in time, for in fossilized cities the oldest levels of human occupation tended to be covered with a series of more recent, superimposed structures built in the same place. Droop was unaware of the concept of archaeological stratification. As had been the case for Pitt Rivers, he determined the stratigraphy of archaeological sites after the fact, on the basis of arbitrarily determined horizontal cuts essentially made to establish the altitude at which important objects were located.

Similarly, the "Wheeler system," which Mortimer Wheeler popularized in the 1950s and above all in the 1960s, was not strictly speaking a method for recording the stratigraphic layers of sites. The "grid method" was primarily intended to facilitate the systematic, three-dimensional rendering of the places where artifacts were discovered. In this respect, it was more an extension of Pitt Rivers' approach to reconstructing terrain than a newly invented stratigraphic method for reconstituting the internal history of sites. In effect, even though the concept of layers had long been recognized by geologists and prehistorians, it was not readily accepted as a means by which to establish the archaeological sedimentation of sites, essentially because archaeology was not really the study of sites, but rather the search for objects or, more generally put, any material creation that could provide information about past civilizations. Layers were simply viewed as the context in which objects were found, not as units of sedimentation that might inform us about a site's material memory. They gave us the period to which objects deposited there belonged, while their formation indicated the general conditions under which objects ended up there: as a tomb or a dwelling was being built or, on the contrary, as they were being razed or abandoned.

Stratigraphic digs were first undertaken in France in the 1950s, by Jean-Jacques Hatt on urban digs in Strasburg and Metz.[34] In the 1960s and 1970s, the method was adopted on archaeological sites for historical periods. Actually, Hatt was simply bringing to post-war France German archaeological techniques dating from the 1930s and 1940s. It was an approach to interpreting terrain that linked archaeological strata with specific historical events, such as wartime destruction or conflagrations that were attested to in historical sources for Antiquity. Once again, terrain was simply seen as contextual evidence of a history that was external to it, rather than the product of a site's internal history, which is to say its memory.

At roughly the same time, André Leroi-Gourhan became one of the first to apply a meticulous, layer-by-layer technique to prehistorical digs in an attempt to uncover the surface of each of the stratigraphic layers. But, here again, the main focus was not the process by which a particular accumulation of artifacts had made its way into the terrain. Leroi-Gourhan was primarily interested in exposing, over as great an area as possible, the location of vestiges "as they had been seen by men in times gone by."[35] His idea of "ethnographic digs" was thus not that different from that of the open-area digs that dated back to the vast archaeological projects of the 1930s. This type of dig sought foremost to reveal the way living spaces or cemeteries, for example, were organized, which, it was thought, could provide direct information about the culture, techniques, and social fabric of given civilizations.

The range had expanded—archaeologists were now looking to grasp the whole of archaeological structures that lay buried in the ground—but the way the land was perceived had hardly changed at all. The approach to stripping it down, which is the process of a dig, had clearly become more meticulous, but that was because it had become obvious that the bits of objects buried in the earth were linked to each other spatially. The principal focus of archaeological fieldwork continued to be objects; the land itself was not seen to hold archaeological memory. Digs still aimed at extracting a bit of the past, which is simply not possible. No more so in the ground than in man's psyche, memories are not pictures of the past, neatly piled on top of each other like so many pages in a book that we need only tear out, one by one, to read them.

As they were first developed in the 19th century, archaeology and the study of prehistory played an important role in the creation of a new universal history, but even that was rear-guard action. The discovery in the 1860s of an age of human history that reached far back—which is to say, an age that history had been unaware of—opened an enormous breach that had to be closed at once. This "dark abyss of time," to use Buffon's expression, had to be filled in with whatever lay at hand. Scholars quickly understood that this seemingly bottomless pit reduced recorded history to an extraordinarily brief episode of the history of mankind and, above all, that it made what was now a lesser history into a layer of relative knowledge hovering above a chasm of unknowns. The history of mankind's progress now looked like one "without a past"; recorded history was just that of mankind's most recent moments. Awareness of the depths of prehistoric time thus opened a widening fissure within the very mass of history, which would henceforth have to be divided into a history—in the conventional sense of the term—of a past that we were conscious of, so to speak, and one that we were not. The latter was the province of prehistory and, in general terms, archaeology. Digs underscored this divorce to the point of making it irrevocable. The more we dug, the more we realized how thoroughly distinct are the history of the events that marked the great civilizations of the past, on the one hand, and the memory constituted by the archaeological manifestations recorded in archaeological sites, on the other. This is the disconnect between *history* and *memory*, generated by the development of archaeology, that Sigmund Freud had sensed following Schliemann's discovery of Troy and Mycenae. From that point on, archaeology became a discipline that brought to the surface vestiges of mankind's subconscious past, subconscious in the sense that it lay buried and had been forgotten.

Clearly, archaeology is not a historical discipline like any other. The history of archaeological research techniques, which we have just looked at briefly, shows that the picture of the material past of human societies that we are able to put together is basically a hybrid made of archaeological matter—as archaeologists discover them at a particular time—and the way these artifacts are perceived by archaeologists. More precisely still, the way in which archaeological vestiges are apprehended depends both on context (that is, the way artifacts are revealed to us physically) and on our reading, at the time, of the material world about us (reality).

This is a naturally cumulative process. The constant addition of discoveries necessitates new, updated interpretations of archaeological matter while, conversely, the way we perceive reality leads us to seek evidence of new phenomena in artifacts. For example, the growing environmental concerns of the last ten years have coincided with new research on man's relation to the environment in archaeological times. And our reading of vestiges of the past is not only hybrid, but also unstable and far more dependent on filiation and heritage than on the accuracy of our observations. Like past material creations themselves, archaeological creations pass on a memory that is constantly being reshaped over the discontinuous time of history. We must now turn our attention to artifacts and see how the mechanisms of transmission and assimilation function within matter.

CHAPTER 6

Notes

1. Zalmen Gradowski, letter in Yiddish, dated September 6, 1944, written in Auschwitz-Birkenau, in Zalmen Gradowski et al., "*Megilot Auschwitz. Les Rouleaux d'Auschwitz (I)*" *Revue d'histoire de la Shoah*, 171, 2001), p. 67.

2. Or "Megilot Auschwitz." Gradowski, *Megilot Auschwitz*.

3. Fragment entitled T.N.Ts.B.H. (for the phrase *Tchi nishmati tsura bitsor hakhayim* that appears on Jewish tombstones and that means: "May my soul be bound to the scroll of the living," and which, in the present case, can be read in the plural *Tihyou nishmotum tsurot bitsrorhakhayim*, "May their souls be bound to the scroll of the living") dated October 10, 1944, written by Zalmen Lewental at Auschwitz-Birkenau. Gradowski, *Megilot Auschwitz*, p. 124. The fragment was found in 1962 in a bottle buried near Crematorium III.

4. Statement by Alter Szmul Fajnzylberg quoted in Teresa Swiebocka, *Auschwitz, A History in Photographs* (Bloomington: Indiana University Press, 1999).

5. Dated January 3, 1945, written in Auschwitz-Birkenau, in Yiddish. *Yehi rotsn milifneikho, eyno shoymea kol bekhies* means "May it be Your will that people shall hear the voice of our tears." (Anonymous, *Revue d'histoire de la Shoah*, 171, 2001), p. 167.

6. Benjamin, *Selected Writings*, vol. 4, p. 390.

7. Marcel Proust, *Remembrance of Things Past*, trans. by C. K. Scott Moncrieff and Terence Kilmartin (New York: Random House, 1981), p. 773.

8. Michel de Certeau, *L'Écriture de l'histoire* (Paris: Gallimard, 1975), p. 125.

9. Certeau, *L'Écriture*, p. 125.

10. Certeau, *Histoire et psychanalyse. Entre science et fiction* (Paris: Gallimard, 1987), pp. 85–88.

11. Certeau, *Histoire et psychanalyse*, p. 87.

12. Alain Schnapp, *La Conquête du passé. Aux origines de l'archéologie* (Paris: Carré, 1993), pp. 148–50.

13. Schnapp, *La Conquête*, pp. 144–45.

14. Paolo Rossi, *The Dark Abyss of Time. The History of the Earth and the History of Nations from Hooke to Vico* (Chicago: University of Chicago Press, 1984), pp. 3–6.

15. Éric Buffetaut, *Histoire de la Paléontologie* (Paris: Presses universitaires de France, 1998), p. 24.

16. Bernard de Montfaucon, preface to *Antiquity explained, and represented in sculptures, By the Learned Father Montfaucon*, trans. David Humphreys (London: J. Jonson and J. Watts, 1721).

17. Markus Castor, "Caylus et le cercle artistique parisien," in Irène Aghion, *Catalogue of the exhibit of the Bibliothèque Nationale's Department of coins, medallions and antiquities* (Paris: Institut national de l'histoire de l'art, 2002).

18. Anne Claude Philippe Caylus, *Recueil d'antiquités égyptiennes, grecques, étrusques et romaines,* 7 volumes (Paris: Desant & Saillant, 1752–1767).

19. Annette Laming-Emperaire, *Origines de l'archéologie préhistorique en France. Des superstitions médiévales à la découverte de l'homme fossile* (Paris: Picard, 1964), pp. 167–75; Claudine Cohen and Jean-Jacques Hublin, *Boucher de Perthes. Les Origines romantiques de la préhistoire* (Paris: Belin, 1989), pp. 201–21.

20. Camille Jullian, "Les Fouilles de Glozel," in *Comptes rendus de l'Académie des Inscriptions et Belles Lettres* (Paris: Picard, 1926); Salomon Reinach, "Les Fouilles de Glozel," in *Comptes rendus de l'Académie des Inscriptions et Belles Lettres* (Paris: Picard, 1926), pp. 203–5, 207, 218–22; (Paris: Picard, 1927), pp. 197–8, 201–3; Salomon Reinach, *Glozel, la découverte, la controverse, les enseignements* (Paris: Krie, 1928); Émile Fradin, *Glozel et ma vie* (Paris: Laffont, 1979).

21. Félicien de Saulcy, "Guerre des Helvètes. Première Campagne de César," in *Revue archéologique*, 2 (1860), p. 337.

22. C. Rossignol, *Découverte d'armes gauloises et romaines réunies, faites dans la plaine des Laumes, sous les murs d'Alise (Côte-d'Or),* 19-page manuscript (Saint-Germain-en-Laye, Archives of the National Museum of Archaeology).

23. C. Rossignol, "Rapport sur les fouilles faites sur les Chaumes d'Auvenay, à une demi-lieue d'Ivry," in *Mémoires de la Commission des Antiquités du Département de la Côte d'Or*, 2 (1842–1846), p. 80.

24. Alexandre Bertrand, "Les Tombelles d'Auvenay (Côte d'Or)," in *Revue archéologique* IV (1861).

25. Laurent Olivier, "Nouvelles recherches sur le tumulus à tombe à char de La Butte à Sainte-Colombe-sur-Seine (Côte-d'Or)," in *Antiquités nationales*, 32 (2000), p. 114.

26. André Leroi-Gourhan, *Les Fouilles préhistoriques (techniques et méthodes).* (Paris: A. et J. Picard, 1950).

27. Alain Schnapp, *L'Archéologie d'aujourd'hui* (Paris: Hachette, 1980); Jean-Paul Demoule, et al., *Guide des méthodes de l'archéologie* (Paris: La Découverte, 2002).

28. One of the very first manuals directed at people doing field research was Jules Bleicher and Jules Beaupré's *Guide pour des recherches archéologiques (époques préhistorique, gallo-romaine et mérovingienne) dans l'Est de la France. Belfort, Doubs, Haute-Saône, Meurthe-et-Moselle, Meuse, Vosges* (Nancy: Crépin-Leblond, 1896). Most of the manuals that followed were published after the turn of the century.

29. Joseph Déchelette, *Manuel d'archéologie préhistorique, celtique et gallo-romaine, I, Archéologie préhistorique* (Paris: Picard, 1908); Joseph Déchelette, *Manuel d'archéologie préhistorique, celtique et gallo-romaine, II, Archéologie celtique ou protohistorique, Âge du Bronze* (Paris: Picard, 1910); Joseph Déchelette, *Manuel d'archéologie préhistorique, celtique et gallo-romaine, II, Archéologie celtique ou protohistorique 2,*

Premier Âge du Fer ou époque de Hallstatt (Paris: Picard, 1913); Joseph Déchelette, *Manuel d'archéologie préhistorique, celtique et gallo-romaine, II, Archéologie celtique ou protohistorique 3, Second Âge du Fer ou époque de la Tène* (Paris: Picard, 1914); Albert Grenier, *Manuel d'archéologie gallo-romaine, I, Généralités et travaux militaires* (Paris: Picard, 1931); Albert Grenier, *Manuel d'archéologie gallo-romaine, II, L'Archéologie du sol (vol. 1, Les Routes; vol. 2, L'Occupation du sol et index des deux volumes)* (Paris: Picard, 1934); Albert Grenier, *Manuel d'archéologie gallo-romaine, III, L'Architecture (vol. 1, Notions préliminaires. L'urbanisme. Les monuments de la vie politique: capitole, forum, temple, basilique; vol. 2, Ludi et Circenses: théâtre, amphithéâtre, cirque)* (Paris: Picard, 1958); Albert Grenier, *Manuel d'archéologie gallo-romaine, IV, Les Monuments des eaux (vol. 1, Aqueducs, thermes; vol. 2, Villes d'eau, sanctuaires de l'eau)* (Paris: Picard, 1960).

30. Michael Thompson, *General Pitt Rivers: Evolution and Archaeology in the Nineteenth Century* (Bradford-on-Avon: Moonraker Press, 1977); Mark Bowden, *The Life and Archaeological Work of Lieutenant-General Augustus Henry Lane Fox Pitt Rivers, DCL, FRS, FSA* (Cambridge: Cambridge University Press, 1991).

31. Mortimer Wheeler, *Archaeology from the Earth* (Oxford: Oxford University Press, 1954).

32. P. J. Droop, *Archaeological Excavation* (Cambridge: Cambridge University Press, 1915).

33. Droop, *Archaeological Excavation,* VIII.

34. Jean-Jacques Hatt, "Pour connaître le passé des villes gallo-romaines. La méthode stratigraphique des fouilles (résultats obtenus à Strasbourg), in *Revue archéologique de l'Est*, II (1951).

35. Leroi-Gourhan, *Fouilles préhistoriques.*

CHAPTER 7

PALIMPSESTS AND MEMORY OBJECTS

In the Middle Ages, a paucity of parchment often led monks to scratch off what was inscribed on manuscripts so they could write something new in its place. We call these superimposed layers of writing *palimpsests.*[1] Etymologically, palimpsest designated the material on which different layers of writing were inscribed, with each new layer contributing to the obliteration of those that preceded it. Since the early 1980s, the term has come to designate the effect created by superimposed archaeological layers or, on the scale of sites or landscapes, the superimposition of areas containing archaeological remains.[2] At the end of the 1970s, aerial photography began producing pictures, often graphically stunning, of archaeological landscapes made of superimposed structures of different kinds, dating from different periods. Viewed from above, the tracings of leveled foundations and filled-in trenches appeared as large light or dark linear shapes that seemed engraved upon the earth's crust. At the same time, the development of large-scale digs, especially in cities, was revealing the structural complexity of sites occupied over long spans of time, marked notably by alternating moments of construction and destruction, frequently separated by periods of abandonment or archaeological inactivity. As defined by archaeologists, the term *palimpsest* thus came to represent a cumulative phenomenon of sequences of human occupation (that produced archaeological remains), thus recording memory in the land in the form of superimposed, stratigraphic layers.

There are actually several kinds of archaeological palimpsests. There are those in which the result of the process of superimposition is clearly visible, such as archaeological structures from different phases of human occupation that have accumulated on a single site, or the superimposed strata of ground on which dwellings were rebuilt over long periods of time. Stratigraphic layers themselves constitute a kind of palimpsest in that their gradual accumulation generally involves alternating periods of activity and inactivity. Some sequences are marked by a process during which erosion, in particular, deposits layers of sediment in which man has placed structures. At other times, erosion or human activity has wiped away or reduced sediment or structures. This is especially the case for ditches or moats dug on previously occupied sites, or for ruins or even foundations that were pillaged for building materials. Thus, what appears to us in the field today as a continuous build-up of strata or of the remains of structures is actually the result of a fundamentally discontinuous history whose "negative" part, so to speak, is invisible.

The process of discontinuous recording is also at work in the creation of archaeological strata on a much smaller scale. Each archaeological layer that appears to us as an individual unit of deposited sediment has generally formed gradually—here too, with alternating periods of activity and inactivity that are not apparent in their internal structure. The artifacts or remnants of artifacts they might hold are usually deposited in a secondary position, meaning that they belonged to periods of occupation that predate the layers in which they are located. These archaeological remains accumulate over periods of time of various lengths and are the result of repeated sequences of deposition, alternating with periods of inactivity, in a single place, only the end-product of which comes down to us. That is exactly what ground with a dwelling standing on it is. What appears to be a single construction (a hearth, a workplace) is actually an accumulation of physical changes most of whose stages are no longer visible. The changes are fundamentally tied to repeated cycles of utilization, with each new cycle partially eliminating the material remains of earlier ones. For example, before coming down to us in its final form, the site of the hearth we have been working on had been a succession of dozens or even hundreds of hearths, each one used for a similar purpose, but each one also unique in what it was like in its time.

The situation is similar for objects, regardless of whether they have remained where they were initially left or have been deposited in a secondary position. Far too often, we forget that a grouping of artifacts found in one place (what is left on the ground when things are made or consumed, for example, pottery fragments, bones, or discarded matter) is the result of material transformations that accumulated over time. Within this process, each new cycle replaced a previous one, either eliminating or altering what remained. Whether it be producing matter or reworking it to create other materials or end-products, it was always a new operation beginning again. Erosion, recycling, repair, alteration, destruction, and abandonment are direct manifestations of the repeated processes of material transformations that are evidenced in vestiges of archaeological activity. The existence of these cycles—known as *chaînes opératoires*—is obviously less apparent when artifacts have been reworked or deposited in layers they don't belong to. Nonetheless, because this information remains partially recorded in the physical transformation of objects, it is never completely lost. Even when reshaped as embankments or fill, or in levels of erosion, vestiges of human activity retain the memory of their former state and are, in this respect, very much like palimpsests.

An assemblage of objects—such as those found in graves—and the objects themselves also constitute kinds of archaeological palimpsests, even when the superimposition of the various states of their existence is not apparent. In the 1980s, work was carried out in Hochdorf (Baden-Württemburg) on an extraordinary find, a "princely" Celtic tomb dating from the 6th century BCE. It contained opulent grave goods, some of which had been reworked for the purpose of being placed inside.[3] For example, parts of the princely regalia—a large embossed bronze-plated belt, his dagger and shoes—had been completely coated in gold, while other objects had been made specifically to be placed beside the prince, such as a pair of gold fibulae for attaching his garments. In effect, the objects located beside the body at the moment the tomb was sealed had actually been made at different times. There were older objects that had been handed down to him (such as, it would seem, the large bronze bench on which he lay), personal possessions (his attire and equipment), and pieces that had been created (or transformed) for him after his death. Thus, the Hochdorf tomb provided information of a sequential kind about the various states and uses of the objects that ended up alongside the body.[4]

Like a monk's roll of parchment, any object can become a palimpsest. There lies recorded within the sequence of physical transformations superimposed upon an object's initial state—through disfigurement, transformation, and deterioration—the memory of a succession of states, each of which had a particular identity and served a particular purpose. As a case in point, today's museums solemnly exhibit objects and tools that were originally intended to be used, not put on display. In a general way, all archaeological matter is at bottom a palimpsest. And this is so because a palimpsest—that is, what results from superimposition, reprising, and repetition—represents the signature workings of memory.

Memory Objects

In fact, rather than speaking of palimpsests in regard to the memory recorded in archaeological remains—that is, in material structures—we would be better served to speak of *memory objects*. A memory object is one in which time is inscribed, or better stated, a material entity in which the memory of a moment in time is recorded. We would do well to reflect a bit on this property that allows us to understand why, fundamentally, recording these "moments of time" in matter is inextricably tied to the effacement—or more exactly, the deterioration—of possible prior or subsequent recordings. The essential feature of a memory object lies not so much in the recording of how the present physically alters matter, as in how the stages of deterioration are recorded, or if you will, memorized. In effect, because changes are preserved in the matter of archaeological objects, they continue to bear witness to their older, now non-existent states. This occurs because all memory objects—be they actual artifacts or the machines we use today to record memory, such as cameras, video-recorders, and computers—share a common trait: Their capacity for recording memory is directly linked to the fact that their sensitivity (that is, the moment when something from the present is inscribed in them) is temporary, both preceded and followed by states of insensitivity or inactivity.

We can get a good sense of the phenomenon by looking at the way photographic images are recorded. In order for film to record an image, it has to have been kept out of the light before exposure; then, once the image is captured, kept from exposure to light until the negative is permanently fixed. Were the light-sensitive medium not protected before and after exposure, it would be impossible to get any image at all. We would end up with a negative so over-exposed as to be opaque. All of which leads to a paradox that casts new light on those memory objects that are archaeological vestiges: One of the basic requirements for imprinting is intermittence, that is, discontinuity. In other words, for something to be imprinted, the medium on which it is recorded—whether photographic film or the ground—has to have been insensitive to what took place both prior and subsequent to impression. Having disappeared and having been forgotten or erased are inseparable from imprinting. We are able to read palimpsests precisely because the discrete phases of inscription are separated from each other by long periods during which absolutely nothing was inscribed. These voids, these vacant periods when time was recorded elsewhere in other forms, allow us to see that the layers of the palimpsest belong to different moments in time. Without them, we would have nothing but a surface filled with traces: an empty, uniform space.

Thus, to imprint time in matter, certain very precise conditions have to be met; if they are not, memory cannot be properly recorded. To this point, we have focused on the process of imprinting rather than on the actual information that is recorded, but it too has to satisfy certain conditions if it is to inform us about the time with which it is associated. The first of these is this: For memory of any kind to be recorded, it must be peculiar to the moment when it is inscribed in matter; that is, it has to have been totally different at every other moment in time. Stated otherwise, if it is exactly the same thing that is inscribed each time in archaeological matter—the same shape of an object, the same type of structure, matter in the same state—there can be no recording of time. All we would see is something that has remained as it was before, much as the crater-pocked surface of the moon appears to us to have been this way for billions of years. Once again, we find ourselves dealing with intermittence and its corollary, repetition. For if each inscription in matter has to be unique if we are to read it, then only one particular moment, captured in one particular conformation, can be recorded to the exclusion of all others. The workings of memory are indeed paradoxical, for the very process of imprinting memory is directly linked to what has been forgotten or has vanished. In fact, as must be clear, the existence of a memory is dependent on absence, on what has not been recorded. At bottom, a memory only exists because of the past that has been lost. For history to have been inscribed in matter, something had to keep on being repeated in the same place; for a palimpsest to tell us something about the history of the writing on the parchment, monk copyists of the Middle Ages had to produce an entirely new document. The internal history of a cemetery can be visible to us only if at different times, in ways both similar and distinct, dead bodies continued to be buried in the ground.

Repetition and Intermittence: From Lyell to Darwin

Something that is repeated, that begins anew, always in the same place: Therein lies the key to unlocking the arrested time of the histories of History. Something that keeps coming back, always new, always there; something tiny simultaneously building the gigantic mass of the world in an infinity of places, and over the full range of scales—that was the tool needed to break through the surface of historical time that was frozen into periods, and to allow the fluid time of history to flow forth as construction. The first men to find this key were geologists, in the first half of the 19th century.

It was the British geologist Charles Lyell's concept of Deep Geological Time that ripped the lock from historical time. In *Principles of Geology, Being an Attempt to Explain the Former Changes of the Earth's Surface by Reference to Causes now in Operation,*[5] Lyell demonstrated that to know the past *as history*, it was necessary to study the present state of the Earth. Lyell began with a postulate that would become known as uniformitarianism. To be sure, he said in substance, it is impossible to know the past events of the history of the Earth, but we can permit ourselves the assumption that the physical phenomena observable today that are responsible for the changes in the Earth's surface, such as volcanic activity or erosion caused by water, existed in the past and have acted in approximately the same way over the course of time.

Lyell demonstrated that when we apply these causes to the vestiges of the Earth's history that have come down to us as geological formations, we note not only that all geological phenomena can be fully explained by causes at work today, but that we are forced to deal with a chronological framework far broader (measured in tens or hundreds of *millions* of years, and not in tens or hundreds of *thousands* of years[6]) than those used previously. Phenomena observable in material vestiges that seemed radical or sudden had actually come down to us over extraordinarily long periods of time. It had happened very slowly, in a gradual process.[7] As Lyell noted, this new, longer geological time provided a historical explanation for phenomena that, in the traditionally brief or tight time of conventional geology, had appeared extraordinary. Lyell made a point of saying that it was not irrational thinking that led geologists to view certain phenomena as prodigious; that was, in fact, the only logical conclusion to be drawn from the observation of vestiges of events that had supposedly occurred over a span of time that was in fact far too short:

> How fatal every error as to the quantity of time must prove to the introduction of rational views concerning the state of things in former ages, may be conceived by supposing that the annals of the civil and military transactions of a great nation were perused under the impression that they occurred in a period of one hundred instead of two thousand years. Such a portion of history would immediately assume the air of a romance; the events would seem devoid of credibility, and inconsistent with the course of human affairs . . . . Armies and fleets would appear to be assembled only to be destroyed, and cities built only to fall in ruins. There would be the most violent transitions from foreign or intestine war to periods of profound peace, and the works effected during the years of order or tranquility would be alike superhuman in magnitude.[8]

The unimaginable expanse of geological time was an essential point for Darwin, who applied Lyell's demonstration not to geological formations but to animal species. Like Lyell, Darwin began by setting forth a basic postulate: "all the chief laws of paleontology plainly proclaim, as it seems to me, that species have been produced by ordinary generation: old forms having been supplanted by new and improved forms of life, produced by the laws of variation still acting round us."[9] In other words, it is the present, and has always been the present, that is the key to explaining the past. And, Darwin emphasized, it is by projecting the processes at work in the present onto the entire duration of the past that we realize the extent to which the vestiges we work with are not to be taken as witnesses to some past history, but far more so as remains that are by definition scattered and fragmentary:

> [I]t is highly important for us to gain some notion, however imperfect, of the lapse of years. During each of these years, over the whole world, the land and the water has been peopled by hosts of living forms. What an infinite number of generations, which the mind cannot grasp, must have succeeded each other in the long roll of years! Now turn to our richest geological museums, and what a paltry display we behold![10]

The considerable impact that *The Origin of Species*[11] has had since its publication in 1859 has made of Darwin the champion of a Malthusian vision of nature as a struggle for survival in which only the fittest and the most adaptable manage to perpetuate themselves. Actually, the process of natural selection was never at the heart of Darwin's research. Natural selection was a kind of hypothetical solution or rational explanation for a phenomenon that fascinated him throughout his life and which he continued exploring until his death. The natural phenomenon that never ceased amazing Darwin was the ability of miniscule organisms, such as coral, or extremely rudimentary, ordinary creatures like earthworms, gradually to create over time something extraordinarily massive. Coral, for example, eventually forms islands. And, as for the humble little earthworm, to which Darwin would devote his ultimate research,[12] it literally produces the vegetable mold that allows plants to grow and, with them, all creatures that feed on them who, in turn, nourish others. Darwin was fascinated by the way in which nature works over long periods of time, specifically with regard to the plants and animals that are right before our eyes.

His approach can be seen as falling within Lyell's "present-based" perspective on geology in that the construction of the past is actually taking place within the present. The repetition of infinitesimal changes that result in the gigantic accumulations of matter evidenced by geological vestiges is, at bottom, simply a basic property of life, the workings of which are visible in the present, right now, not in fossils that are but a distant product of it. In other words, the past as history can be explained not through what remains of the past—which for us means vestiges and writings—but through what is alive in the present. And the processes that determine the repetitive phenomenon by which time creates nature are the habits of plant and animal species. This is what Darwin studied so painstakingly: the way in which habits regenerated over time could create something new; in other words, the way in which nature's ways generate history.

Beginning in 1837, Darwin compiled his observations on phenomena that could serve as evidence of the *evolution* of species. "It appeared to me," he wrote in his 1876 *Autobiography*, "that [I needed to follow] the example of Lyell in Geology, and [collect] all facts which bore in any way on the variation of animals and plants under domestication and nature."[13] Darwin explained how the publication of *The Origin of Species* had been something of a chance happening. In June 1858, while he was working on a long piece on species, he received a letter from a then little known naturalist working in Southeast Asia, Alfred Russel Wallace, asking him if the conclusions he had reached merited publication. It happened that Wallace's thesis was the very one that Darwin had been working on for years; namely, that from the moment the varieties of a given species appear, they tend to drift away from their original type through repeated reproduction. He had to act fast, lest he lose the scientific benefit of more than twenty years' worth of compiled observations. In this respect, *The Origin of Species*, which was published in November 1859, just eighteen months after he received Wallace's letter, has to be considered a book hastily written in an attempt to protect his findings. Darwin had no time to expand on anything. On the other hand, it was important for him to offer specific arguments in anticipation of the criticism his theory was sure to spark. This is manifestly one of the major reasons why *The Origin of Species* centers on the thesis of natural selection. It offers a plausible explanation for the variation in species and thereby goes beyond Wallace's observations by making them part of an overall demonstration.

The publication of *The Origin of Species* interrupted Darwin's work synthesizing the observations he had been compiling on the variations in species since 1837. Once he did return to the problem, it would be another ten years before he could publish a book on the subject: *The Variation of Animals and Plants under Domestication.*[14] It appeared in 1868, and was actually an expanded version of the first two chapters of *The Origin of Species.* Darwin made especial use of his personal experience raising pigeons to analyze the combined effects of environment (such as food and climate) and habits (such as the special use animals make of certain parts of their bodies) on the transmission of individual characteristics through reproduction. Up through the 1850s, in studying the phenomena of transmission through reproduction in domestic species, Darwin had worked with breeders and horticulturists who selected and reproduced animals and plants. "I soon perceived," he would later say in his *Autobiography*, "that selection was the keystone of man's success in making useful races of animals and plants. But how selection could be applied to organisms living in a state of nature remained for some time a mystery to me."[15] In all events, by the 1840s, his observations had convinced him that all natural species could evolve; that is, that they were in no way immutable.[16] Getting species to evolve: that is what people in stockyards and gardens were doing routinely when they intervened in the reproductive process. But how could nature do this by itself?

Darwin said that the solution to this crucial problem came to him in October 1838, as he was reading Malthus' *An Essay on the Principle of Population*, which he had picked up to take a break from his study of plants and animals.[17] As he noted in his *Autobiography*: "being well prepared to appreciate the struggle for existence which everywhere goes on from long-continued observation of the *habits* of animals and plants, it at once struck me that under these circumstances favorable variations would tend to be preserved, and unfavorable ones to be destroyed. The result of this would be *the formation of new species*." (emphasis added)[18] Malthus' essay provided Darwin with a partial explanation for the basic way nature works, which until that point had escaped him. He reasoned that the artificial selection breeders and horticulturalists engaged in with domestic species occurred naturally with species in nature through *demography*. In other words, demographic forces regulate the repetition induced by habit or behavior and the innovation that comes with the reproduction of individuals.

In explaining selection as a process favoring the "success" of species, Malthus brought to light another problem whose importance Darwin admitted he had not fully appreciated. Not only do species evolve into varieties through "natural selection," but also there exists, more remarkably still, a "tendency in organic beings descended from the same stock to diverge in character as they become modified."[19] But why do things evolve in this particular way? Why does nature alter a successful formula? And, once the process has been initiated, why does it continue at an accelerating rate? Here again, Darwin discovered that the key to the phenomenon resided in that triangular relationship between repetition, innovation, and demography. "The modified offspring of all dominant and increasing forms tend to become adapted to many and highly diversified places in the economy of nature."[20]

Lyell and Darwin opened a new field that made it possible not so much to record the history of nature as to understand how the history of the world takes shape. In a totally unexpected fashion, it became clear that the rudimentary, the infinitesimally small, allowed us to account for material creations as massive as geological remains. It wasn't extraordinarily powerful forces that had created accumulations of gigantic rocks, just as it wasn't violent rifts that were responsible for stark dissimilarities in a succession of strata or fossil species. It was the work of something minute that had been accumulating continually, a reiterative force that had been relentlessly transforming the material of the present. In its form, the history of the Earth was a palimpsest; in its functioning, it was memory incarnated in the materiality of things and beings, in their conformation and their behavior. In Lyell's work, as in Darwin's, it is striking to note the radical degree to which their very concrete observations about rocks, animals, and plants have altered the perspective from which we gain access to history. The past is no longer the exclusive province of history, as conventional history, history made of stories, would have it. And this is so because, even though radically different from the present in its conformation, *the past is not essentially distinct from it.* It is the product of the same processes that generate the present and that are at work right before our eyes. It is now the present that allows us to understand the past, in a way, intimately. As a consequence, history—as knowledge of the past—has been transformed. It is no longer the story of a succession of time periods of the past. It has become a heretofore unknown discipline given to viewing the past as the ongoing accumulation of memory. The work of Lyell and Darwin gives us a sense of how archaeology could have changed had it undergone a revolution similar to those that occurred in geology and paleontology.

We cannot leave these two men without mentioning the way in which each of them used theory, not to relate the past, but to inform it. Once again, the order of things was reversed. Darwin, in particular, did not draw his theory of natural selection from paleontological vestiges, even though these were to constitute the primary field to which he would apply his vision of natural history. He rather drew his theory entirely from the present, and then superimposed it on the past. That is an approach that archaeologists would deem wholly inappropriate, for their attempts at theorization have until now focused on accounting for the conformation of artifacts as they appear when found.[21] Darwin didn't even bother to address that issue. His "theory of evolution" was a tool he used to depict the past, not as a fixed moment in history—a unique time frame with its own identity—but as uninterrupted movement, a process that has no specific place in time, or rather that is always "here and now." And it is precisely this approach to the past, as memory and not as history, that makes it possible for us to evaluate what the material vestiges of the past mean in relation to the past from which they come to us. As Darwin said in *The Origin of Species*, "I do not pretend that I should ever have suspected how poor a record of the mutations of life, the best preserved geological section presented, had not the difficulty of our not discovering innumerable transitional links between the species which appeared at the commencement and close of each formation, pressed so hardly on my theory."[22]

Here, theory gave form to the reality of artifacts, not the other way around. It brought to the fore something that was effectively not intended to be seen; namely, the incomplete, truncated nature of the information recorded in fossils, which was the only information there was. We would never come upon, or work with, anything other than bits of the past. We too know in our own way that artifacts are incomplete. Darwin gave us to understand why: because intermittence, which, as we have seen, is central to palimpsests, is an essential property of the formation of fossils, in other words, of the sequential recordings of manifestations of the past. "It would seem," Darwin wrote in *The Origin of Species*, "that each separate formation, like the whole pile of formations in any country, has generally been intermittent in its accumulation." He then went on to say that, "We do not make due allowance for the enormous intervals of time, which have probably elapsed between our consecutive formations, longer perhaps in some cases than the time required for the accumulation of each formation."[23] We get a false picture of time from what appears to us, in rock or in the soil, as a sequence of vestiges—animal fossils, remains of structures or objects—lodged in a continuous series of layers. When we do manage to determine (or approximate) the date when these artifacts made their way into the ground, their deposition appears fundamentally episodic, almost anecdotal, because the broader perspective of the duration of a given site is largely dominated by periods in which no archaeological or paleontological remains were left.

The Memory of Matter: The Contribution of Psychoanalysis

Following Lyell, Darwin showed us that the remains of the past, as "fossils" or "palimpsests," were essentially the product of memory rather than of history. Strictly speaking, past events do not produce testimony or matter that one need only collect to be able to reconstruct the history they were a part of. It is rather, more prosaically, the process of reproduction, of reinscription in matter—a process common to everything that lives—that creates history, which has neither beginning nor end, and, manifestly, is recorded only intermittently; and even this intermittent recording is grossly incomplete. This, in truth, is what we deal with: insignificant bits of the past. And with this realization, the concept of history as the possible story of what happened in prehistorical times comes tumbling down.

Is this to say, then, that we must abandon all hope of saying anything at all about the past? We need to go back to palimpsests in their fundamental relation to memory to understand the kind of long-range historical construction of the process of "reimprinting" or being "brought back into play" that reiteration or reproduction generates. Interestingly, Sigmund Freud saw a connection between the structure of the unconscious—the psychic substance in which memory is imprinted—and the way palimpsests are constituted. In his use of an archaeological metaphor, Freud sought to bring to light two essential implications of memory that, in turn, inform archaeology in the discipline's particular relation to "memory objects," which is what palimpsests are:

- On the one hand, the imprinting of layers of memory in the psyche does not function as simple superimposition, in that what already exists is transformed or deformed by what is added. The point is less obvious than it may appear. It means that each new episode in the construction of memory effectively rewrites or reinterprets what was previously recorded. For archaeology, the phenomenon is especially clear with regard to processes of occupation, or reoccupation. Each new stage in the history of a site, which is marked by the creation of new structures, involves either reinscription, in the present, of structures that have disappeared (when their form is reproduced in new structures) or retranscription, when older forms are "translated" into new conformations.[24]

- On the other hand, initial or earlier events that are inscribed in memory are accessible only through their subsequent rewriting(s), and not as they actually were, if that phrase actually means anything, given the constant transformation. Thus, the character of events of the past is essentially established through deferred action, to use Freud's expression. Similarly, the subsequent layers that are added to older formations are of paradoxical significance in that, although they offer access to past events, they do so through a distorted transcription of them. Here again, human occupation of sites over vast spans of time offers explicit examples of this phenomenon. Monuments, cities, and even landscapes evolve through a process of constant change, in which the original function of places continues to be carried forward although it is altered to respond to new needs. In the same way, although the nature of structures is, as a rule, clearly established at the outset (as with groupings of monuments), the overall nature of a site can either be reaffirmed or, conversely, undone in the course of later habitations. More precisely stated, the initial nature of a site is established only "after-the-fact," either through confirmation (the addition of new structures of the same kind) or nullification (abandonment).

Following in Freud's footsteps, Jacques Lacan took the palimpsest image of the unconscious further by focusing attention on the discontinuity that results from the intermittent process of accumulation, which is how memory is constituted. Like the earlier, effaced writings on palimpsests or the vestiges of ancient sites that lie buried beneath present-day habitations, the unconscious, for Lacan, was a "deleted chapter of a person's life," the place where the continuity of the historical meaning, as it were, of memory had been interrupted. Lacan noted in particular that the cumulative structure of memory was fundamentally discontinuous, because the layers that accumulate over time are not necessarily linked to each other. That is a problem we are very familiar with in archaeology. On a given site, we frequently encounter cumulative sequences of different modes of occupation whose initial function has become unintelligible. For example, abandoned parts of rural structures that were used throughout the Roman occupation of Western Europe might become cemeteries in the Early Middle Ages. The changes are never gratuitous. As we have already seen, the effect of discontinuity or intermittence is indicative of the reiterative process of evolution. The way in which the present effaces the past was of interest to Lacan because of their reciprocal relation; that is, the way in which the unconscious—buried prehistory—remains accessible in spite of the conscious—what we would call the living history of ongoing sites:

"[The subconscious is a repressed language] that never disappears," he said in an interview with Gilles Lapouge published in *Le Figaro Magazine* in 1966. "It is present within us, even if we don't have access to it, and it constantly resurfaces through the cracks of our consciousness. It is the mechanism that Freud referred to as the 'resurfacings of the repressed' whereby there intervenes, beneath the clear voice of our consciousness, another, urging, repetitive voice recounting serious tales of our prehistory, which we don't understand . . . . I think we can use the image of the palimpsest, . . . those manuscripts on which the initial writing was erased to allow for a new writing. That's it, a palimpsest: there are two writings to be read, one of which reappears only where the other one has faded, but which is not related to the newer one and whose voice we will not hear so long as we fail to identify its structure."[25]

These are highly significant observations. They provide a phenomenology of the surviving traces of the entombed past. Whether it be buried, disfigured, or unknown to us, the distant past whose memory has been lost continues to manifest itself in the present, "through," as it were, all of the subsequent accumulations. It comes back constantly, Lacan said, through the cracks and the fading, a ubiquitous presence, but one that we are unable to recognize. Recent research on landscape archaeology, which Gérard Chouquer calls archaeo-geography, offers a host of examples of this dynamic transmission of a buried past.[26] One of the most readable sites is located near Orange: the Malalones site in Pierrelatte, where a network of ditches reproduces the Roman organization of the land.[27] Archaeology demonstrates here that an ancient (for us) occurrence (the system of land division dating from the beginning of Roman occupation) created an initial archaeological structure (a network of orthogonal demarcations based on Roman metric units) that over the course of time has been reproduced on a number of occasions and that still "survives" today in the boundaries of tracts of land marked by hedges and rows of trees.

The digs revealed that bringing the Roman system "back in play" was reiterative (a series of ditches dug in the same place) and intermittent (there were manifestly long periods of time during which the ditches, which had been completely filled in, were not being used). It is fair to say, then, that the Roman system of land division, even though buried and forgotten, continues to inform the present; that the Roman past created a "potential" that could be brought back into play again and again and whose "urging, repetitive voice," to use Lacan's expression, can still be heard. In effect, it is in the "creases," if you will, of the present-day landscape—in details as seemingly insignificant as hedges and rows of trees—that we discern the ubiquitous, persisting traces of the buried past, a past that, even though it is "repressed," keeps coming back today, just as it has been coming back for nearly two millennia.

Having seen the stratification involved, in other words, the structure of this palimpsest, we now understand how a present-day landscape can bear the imprint of a Roman past to which nothing is actually connected. Lacan was right in saying that, for us to be able to decipher the presence of this past, we have to have identified its structure (in this case, the typically Roman land division). Failing that, we see only an ordinary rural landscape, in which everything seems to be happening in the present, in a "conscious state," to use psychoanalytical terminology. Now that we know how to identify traces of the past surviving in the present, we understand what Lacan meant by fading, when he said that the subconscious past emerges when present consciousness fades. In Pierrelatte, the persisting traces of Roman land division are present in places where the current one, incapable of imposing its own distinctive mark, has reproduced a past from which, manifestly, it had not been able to free itself.

The subconscious past has a structure of its own that has to be deciphered to be seen. (As Lacan's well-known observation had it: The subconscious is structured like language.) Approaching it from the outside, as it were, as something from some completed past that now lies buried, does not allow us to learn what it is made of or what it is saying. Psychoanalysis teaches us that the past has meaning in the present, in fact that it has meaning only in its relation to the present: There can be no constitution of the present without it. Like the Roman network inscribed in a present-day landscape, the subconscious past is like writing that we can see but that we do not know how to read. Its manifestations are reminiscent of the Egyptian hieroglyphs before they were deciphered by Champollion, a language, as Lacan pointed out, that is at once present and lost. Archaeologists are familiar with these languages; they can recognize them, but they can't decipher them. The inscriptions speak—recalling events, naming people, invoking Gods—but because of our inability to read them, no one can hear them. To consider them merely juxtaposed depictions—recognizing that this one looks like a bird, or that this other one is undeniably a person—would guarantee that they will not be understood, compounding ignorance with incomprehension. Deciphering the characters of a written language involves recognizing the relationship between them. This is the theory that is missing in archaeology, and it is precisely this kind of knowledge of the past that Lyell and Darwin attempted to develop. Could it be that archaeology, because of its focus on vestiges, is closer to the natural sciences, and to biology in particular, than we think?

Notes

1. From the Greek, *palimpsestos.*

2. In particular, in the works of British archaeologists Geoff Bailey ("Concepts, Time-scales and Explanations in Economic Prehistory," in A. N. Sheridan and Geoff Bailey, *Economic Archaeology* (Oxford: British Archaeological Reports, International Series, 1981), pp. 109–10, and L. Patrick "Is There an Archaeological Record?" in *Advances in Archaeological Method and Theory*, 8 (1985).

3. Jörg Biel, *Der Keltenfürst von Hochdorf* (Stuttgart: Konrad Theiss, 1985).

4. Laurent Olivier, "The Hochdorf Princely Grave and the Question of the Nature of Archaeological Funerary Assemblages," in Tim Murray, *Time and Archaeology* (London: Routledge, 1999).

5. Charles Lyell, *Principles of Geology. Being an attempt to explain the former changes of the Earth's surface by reference to causes now in operation.* (London: John Murray, 1830).

6. This is what Charles Darwin demonstrated with regard to the "denudation of the Weald" in Great Britain. Charles Darwin, *On the Origin of Species by Means of Natural Selection, or The Preservation of Favoured Races in the Struggle for Life.* (London: John Murray, 1859), ch. 9.

7. The theory that posits uniform rhythms for the Earth's transformations in the present and the past is known as *gradualism.*

8. Lyell, *Principles*, vol. 1, pp. 78–9. http://books.google.com/books?id=dsYQAAAAIAAJ&dq=Principles%20of%20geology%20Lyell%201830&pg=PA78#v=onepage&q&f=false

9. Darwin, *The Origin*, ch. 10.

10. Darwin, *The Origin*, ch. 9.

11. Darwin, *The Origin.*

12. Charles Darwin, *The Formation of Vegetable Mould through the Action of Worms, with Observations on their Habits* (London: John Murray, 1881). In this last of his works to be published during his lifetime, Darwin used material from an article written more than forty years earlier, in 1838.

13. Darwin, Charles, *Autobiography*, online.http://darwin-online.org.uk/content/frameset?itemID=F1497&viewtype=text&pageseq=1 (p. 119).

14. Charles Darwin, *The Variation of Animals and Plants under Domestication* (London: John Murray, 1868).

15. Darwin, *Autobiography*, p. 119.

16. In a letter to his botanist friend Joseph Hooker, on January 11, 1844, Darwin wrote: "I am almost convinced (quite contrary to the opinion I started with) that species are not (it is like confessing a murder) immutable," in *The Correspondence of Charles Darwin*, ed. F. Buckhardt and S. Smith, vol. 2 (Cambridge: Cambridge University Press, 1987), p. 23.

17. Thomas Malthus, *An Essay on the Principle of Population* (Oxford: Oxford University Press, 1993).

18. Darwin, *Autobiography*, p. 119.

19. Darwin, *Autobiography*, p. 119.

20. Darwin, *Autobiography*, p. 119.

21. Researchers have been especially keen to find signs of regularity that would indicate the existence of particular constraints or "laws" (for example, Michael Schiffer and the processes by which archaeological rejects are constituted; Michael Schiffer, *Formation Processes of the Archaeological Record.* (Albuquerque: University of New Mexico Press, 1987).

22. Darwin, *The Origin*, ch. 9.

23. Darwin, *The Origin*, ch. 9.

24. For example, the renowned Stonehenge site in Wessex went through seven major construction phases in the course of which the original large, megalithic circle was both maintained and modified. The first Stonehenge, which dates from the fourth millennium BCE, consisted of a vast ring 100 meters in diameter, probably with wooden superstructures of the Henge type. The second Stonehenge, from approximately the middle of the third millennium, was notable for its double circle of large bluestones in the middle of the original monument. That pattern was replaced by the ring of sandstone blocks with lintels that are still standing today, and which date from the second half of the third millennium. The last stages of reworking go back approximately to the end of the first half of the second millennium BCE.

25. Jacques Lacan, Interview with Gilles Lapouge in *Le Figaro littéraire*, no. 1076 (1966), p. 2.

26. Gérard Chouquer, "Crise et recomposition des objets: les enjeux de l'archéologie," in Gérard Chouquer, "Objets en crise, objets recomposés. Transmissions et transformations des espaces historiques. Enjeux et contours de l'archéogéographie," in *Études rurales*, no. 167–8, 2004.

27. C. Jung, "Morphogenèse, fonctions et évolution de la centuriation B d'Orange et essai de restitution diachronique des paléopaysages du Tricastin (Drôme-Vaucluse)" (Tours: Université de Tours, unpublished doctoral thesis, 2 volumes); Chouquer, "Objets en crise."

CHAPTER 8

A BIOLOGY OF FORMS

Diary of a Mad Man

They are waiting for the dance to begin. "Grave and attentive," they are staring at the box with the little glass and metal eye that the strange white man in heavy black clothes is pointing at them. The little ones, in the front, have not noticed his presence. They are sitting on the ground, "like savages," we might say, playing in the dirt in the pale sunlight. All of the older boys have seen him. They are huddled together in the middle of the group, eyeing somewhat incredulously this stranger who has come to their village, just as they look at us from their world long since left behind, a world that is incredibly plain and dry, a world of earth and stone. They are clearly awed by the man in black, with his hat and his big mustache and the gold watch chain glittering on his vest, but they are not afraid of him. For them, it is as if he were the one who was not completely real. He would like to talk to them, but he doesn't know them, and doesn't speak their language. So he just smiles, with a weary somewhat absent smile. He has come here from Europe, after a two-day trek across the desert, from the farthermost train station in America. His name is Aby Warburg, and he is known to be a German art historian specializing in the Renaissance. He is lost in the deepest reaches of Arizona. He is mad.

Warburg was rich. Everyone in his family was in banking in New York or Hamburg. When they were children, his older brother promised him that he would never have to work, that he would buy him all the books he wanted—and he kept his word. Aby amassed some 80,000 volumes in the course of his life. Warburg was subject to phobia and anxiety. During the preceding winter, in the Mesa Verde gorges in Colorado, he was obsessed with the thought of catching pneumonia[1] because he felt that during the cholera epidemic he "did not hold out in Hamburg, as [his] brother and dear wife's family did."[2] The peculiarity of Warburg's mental illness, which was to surface during World War I, lay not so much in these irrational fears as in the fact that, involuntarily, they literally became reality. In this light, there was already something pathological about the map he drew during his 1895 trip to the Mesa Verde region, with every campsite and every place he stopped to eat clearly marked. Warburg's sketch of the landscape shows a network of tree-lined ravines that look like lungs, and the whole of it suggests a skinned carcass with the ribs showing.[3]

In the spring of 1896, when he went to visit Pueblo villages in Arizona, Warburg wrote that he felt like a "seismograph of the soul," "on the dividing line between cultures," that of his Jewish origins and personal life in northern Germany, and that of his intellectual pursuits that focused on the Italy of Antiquity and the Renaissance. Later, he would say that he felt "driven towards America, the object of a supra-personal duty, in order to experience life there in its polar tension between the instinctive, pagan nature cult and organized intelligence."[4] Intellectually, Warburg knew well that this "tension" or "cleavage" was a symptom of a kind of madness, an "internal, schizoidal contradiction."[5] He actually said so in a lecture he gave in 1923, at the Bellevue Clinic in Kreuzlingen, Switzerland, where Ludwig Binswanger had been treating him for schizophrenia since 1921.[6] The comparison with the seismograph was not mere metaphor for Warburg; it was reality, or in any event, he lived it as if it were during his fits of madness, when life would take hold of him as if he were an object. He could see everything then, could understand everything all at once, because reality—whether real or imagined meant little at that point—passed through him, *within him.* Like the needle of a seismograph jumping nervously up and down as it records earthquakes, Warburg's body and mind reacted to the tension he felt reverberating within him as "instinct," he said, pulled against "inhibition," as the power of "magic" pulled against "destructive logic."[7]

In Warburg's perception of reality, there was nothing resembling an idea or an interpretation, not as it is commonly understood by those of us who have no claim to madness. Describing it through words and pictures was his way of holding it at bay, of keeping the devastating tension at a distance. In fact, when he gave his famous lecture on the Pueblo Indians' "snake dance," in 1923, at the clinic in Kreuzlingen, Warburg made a point of saying that his talk was not to be taken as a research paper presenting scientific conclusions. It was rather "the desperate confessions of someone seeking redemption from a state in which his attempt at spiritual elevation has been arrested by [Warburg later corrected to *in*] the compulsion to be connected through *a real or imaginary* incorporation."[8] Warburg wanted to escape from the unendurable vibrations of the tension that fills reality, and of which the conventional division separating the real from the imagined, truth from fabrication, has absolutely no grasp. His encounter with the Pueblo Indians and their "snake dance" showed him that, in all societies, both present and past, cultural creations are seismographic in nature. Representations—more precisely, images and shapes—are at bottom, he wrote, a "biologically necessary product situated between the practices of religion and art."[9] The zigzag outline of the snake's lightning-like body, which for the Pueblos linked earth and sky, was to Warburg's mind a seismographic form. In Kreuzlingen, even his handwriting, which suddenly began to jump up and down across the page, or which would suddenly be struck through by lightning bolts similar to the zigzag bodies of the Hopi snakes, became seismographic.[10] On August 8, 1923, on the title page of his draft of the "snake dance" lecture, Warburg penned the word "Help!"[11]

Warburg's writings were supposed to be published posthumously by the art historian Ernst Gombrich in one "great work" in several volumes. The work never appeared. Instead, in 1970, Gombrich published *Aby Warburg: An Intellectual Biography.*[12] It was an attempt to give intellectual legitimacy to Warburg's contributions to contemporary art history. To do so, Gombrich eliminated a considerable number of pieces he considered to be extraneous, so that some of Warburg's most important writings remained unpublished for a very long time. It wasn't until 1988, for example, that the *Snake Dance* was published in German.[13] French translations did not begin appearing until almost the turn of the century.[14]

Their publication notwithstanding, it remains clear that we still do not know how to deal with the man's madness. Gombrich transformed Warburg's "seismographic" study of shapes and images into an "iconology" stripped of its pathological, schizoid dimension. The schizophrenic collages that Warburg put together for a collection of Greek, Roman, and Renaissance iconography and his vast, unfinished *Bilderatlas Mnemosyne* project never spawned followers. In truth, only Warburg's tortured, distraught mind was capable of identifying the pertinent details of the forms in figurative art of different periods and styles, which were for him evidence of repeated "seismographic" responses to the primitive tensions to which image-making and shape-making were always subjected.[15] Conversely, the temptation has been great to integrate Warburg's reflections into the thoroughly traditional approaches of the Academy where, normalized at last, they could take their place. Warburg's ideas have to be seen for what they are: pathological and delirious, to be sure, but extraordinarily perspicacious and fecund.

Image as Symptom, Form as Palimpsest

What exactly was Warburg after on his visit to the Pueblo Indians? "I had developed a downright disgust," he would write in Kreuzlingen—while "still under opium"—"with estheticizing art history. The formal contemplation of images—not conceived as a biologically necessary product situated between the practices of religion and art (which I understood only later)—seemed to me to give rise to such a sterile trafficking in words that after my trip to Berlin in the summer of 1896 I tried to switch over to medicine."[16] Images that were the result of a process of mediation (to borrow another of Bruno Latour's phrases) occurring in a field of tension generated dialectically by the opposing poles of representations and practices: that was what Warburg went off to see in the American Southwest. He wanted to observe the Pueblo Indians "snake ritual," a ceremonial group dance designed to make the land fertile through an appeal to living snakes. For Warburg, the dance reproduced the Dionysian processions of Classical Antiquity. He was determined to witness it first hand.

Warburg's belief that the Ancient Greeks and the Pueblos of the late 19th century were "kinsmen,"[17] in spite of the distance in time and space separating them, was heresy for conventional art history, and sacrilege for every other historical discipline based on a historicist approach to the past. It was madness to believe that two distinct periods in time could be related when there could obviously not have been communication between them. It was even more scandalous to claim that two cultures so distant from each other—one "savage" or "primitive," the other, moreover, representing the quintessence of civilization—could belong to the same family when they were so far apart on the line of civilization's development. The postulate had been originally formulated by Johann Joachim Winckelmann:[18] It was history that gave civilizations, or cultures, their identity; they could exist only if they had their own place in time and space. Warburg countered with a radical critique of art history and of traditional history in general. In his view, these were superficial approaches limited to the formal description of works, nothing more than commentary on things said or represented. The conventional approach to the history of civilization was sterile because it was incapable of getting to the meaning of the art of the distant past. It was nothing more than chatter packaged as erudition.

Warburg based this extreme position on a phenomenon that the sequential approach to civilizations tended to ignore: persisting traces. The shapes and motifs found in Renaissance art, for example, had not simply been reproduced or imitated from classical Antiquity. What appeared in 16th-century Christian art, Warburg insisted, were actually reworked representational themes dating back to ancient pagan times. In the process, they had assumed a new existence, a "posthumous life" (*Nachleben*). And this was how cultures, or civilizations, communicated with each other across time and space. They re-evaluated or recomposed shapes they held in common. In effect, they had all been subject, over time, to tensions of a similar kind. Cultural creations were not to be viewed simply as images that were either simple or complex, crude or elaborate, pleasing or unpleasant. They had to be understood as parts of a memory that had developed over the long course of civilization.

Whether this view was the product of Warburg's deranged mind is a moot point. His obsession with identifying in the new the persisting traces of the old touched on an essential aspect of cultural representation or, more generally stated, material manifestations; namely, the idea that these creations were not to be taken as physical evidence but more so as symptoms. By this, Warburg meant that Greek and Roman sculpture, or Renaissance paintings, or even Pueblo Indian rituals, expressed more than just the particular cultural identity of the people or historical period that produced them. It was, moreover, because conventional art history traditionally reduced them to cultural representations of the groups that created them, that the discipline had bound itself in a sterile aesthetic.

Warburg was less interested in what these creations represented than in the purpose they served. For Warburg, the cultural creations of any given period were never gratuitous. He saw them as generated by a field of tension whose opposing poles were "instinct" and "intelligence," "contemplation" and "thought," or more generally, "the cosmos in its natural state" and "civilization." The cultural creations of a given civilization transcended time, and their forms survived because they allowed the societies that produced them to mediate with these opposing forces and reconcile themselves with them. The contexts in which snakes were depicted were the same for the ancient Greeks and Romans as for the Pueblo Indians of the American Southwest, Warburg insisted, not because the snake's cultural significance was the same for each of these cultures, but because the appeal to the figure of the snake was an attempt to respond to the same kind of tension. Herein lies the fundamental reason why Warburg's time with the Pueblo Indians was so profoundly important for him. His encounter with the Pueblos shook his knowledge of Renaissance and Greek and Roman art to the core, and allowed him, he wrote, to "distinctly see the identity, or rather the indestructibility, of primitive man, who remains the same throughout all time."[19]

Image as symptom rather than as physical evidence: Warburg was not interested in form as it appeared—its style and characteristics—but rather in what he called its "heraldic skeleton," which he could see in the ornamental patterns (for there were more than one) on Hopi ceramics. The "background" geometrical patterns representing earth and sky that were linked by thunderbolt-snakes bringing rain were there not so much to depict as to signify. Similarly, replications of this "background," of which there are a number of variations, were intended not so much to represent as to transmit this significance. What is perpetuated over time is not so much a form as the skeleton of a form, just as it was not exactly a composite of symbols—such as those found on a medieval coat of arms—but rather a particular shape represented in a variety of ways.

Because form is symptom, it is unstable. As Warburg noted, it is the result of "a compromise between image and sign, between realistic representation and (pictorial) writing." As in heraldry, this compromise between forms of representation that are so radically opposed necessarily involves a composite of heterogeneous elements. In this regard, according to the anthropologist Carlos Severi, the cultural creations of a civilization are fundamentally chimera, or, better stated, "chimera-objects."[20] Warburg didn't go to the American Southwest, as Tylor had when he journeyed to Mexico,[21] to observe what had survived of the primitive, supposedly aboriginal culture of the Pueblo Indians before it became completely and definitively Westernized. He saw in contemporary Indian culture "contaminated matter," that is, stratified matter: "What originally existed in America," he wrote in regard to American Indians, "has been covered since the end of the 16th century with a layer of Spanish-Catholic confessional teachings whose influence abruptly ceased at the end of the 17th century . . . . Then another layer was added, that of North-American teachings, covering the whole of it."[22] Thus, Warburg was interested in contemporary Amerindian culture as composite matter. He wanted to study not just the additive process of superimposition that had transformed it, but also, and perhaps most of all, the way in which old and new continued to interact within the superimposed cultural "layers." Warburg said that this stratification represented "the most difficult object imaginable" for it was very much "a palimpsest whose text—even if we bring it out—is contaminated."[23]

The various, heterogeneous, contaminated cultural creations of a civilization live on as long as the tension in which they are caught is active. Once these forces are truly overcome—that is, when they have lost their effect—images die out, having lost their *raison d'être*. That was precisely what killed the Indians, or more precisely, what resulted in their acculturation. Warburg said this:

> As I was walking in the streets of San Francisco, I had a vision of the man who had triumphed over the snake cult and the fear of lightning, the latest of the descendants of primitive man and of the gold digger who had wiped out the Indians. It was Uncle Sam with his top hat, walking proudly down the street in front of a circular, neo-classical building. There was an electric wire strung over his hat: Edison's copper snake that had stolen lightning from nature . . . . The telegraph and the telephone have destroyed the cosmos. Mythical thinking and symbolic thinking had struggled to bring a spiritual dimension to man's relation to his environment, and made space a place for contemplation and thought. Instantaneous, electric communication has destroyed that space.[24]

CHAPTER 8

Archaeological Typology: A Biology of Forms

The Italian philosopher Giorgio Agamben said, as Robert Klein had before him,[25] that Aby Warburg's work created a discipline "which, unlike most, exists but has no name." Agamben called it "a science without a name," "an unnamed discipline."[26] What did it involve? Although he did not realize it—Warburg was an art historian, not an archaeologist—he had brought back to life what had been for prehistoric archaeology in the 1870s a stillborn dream: the failed encounter between prehistoric typology and Darwin's theory of evolution. Even back then, prehistorians attempting to reconstruct the evolutionary logic of artifacts had dealt with the emergence of form as a *necessity of a somewhat biological order*, to use that phrase again. What were the forces, they asked, that determined the way the morphological or stylistic characteristics of the material creations of the past had developed? In other words, what had assured the sustainability of such innovations and guaranteed that they would be transmitted over time, from one generation of objects to the next, so to speak?

Drawing on the works of Thomsen and Worsaae, Scandinavian archaeologists were the first to develop a chronology for the cultural creations of prehistoric civilizations, principally those of the Metal Ages. These earliest observers were struck by the manifest filiation in the morphology (or style) of objects transmitted and modified over time. Once created, certain morphological characteristics tended to evolve over time as they were reproduced on different objects or materials. What drove this evolution of archaeological matter? In the early 1870s, the Swedish archaeologist Hans Hildebrand drew a parallel between the morphological evolution of artifacts and the biological evolution of living species that had been reconstructed by paleontologists. More significantly, Hildebrand saw in the connection between prehistory and paleontology the basic principle of the chronological sequencing of prehistoric objects. He believed that the evolution of the technical and stylistic characteristics of objects from the past allowed us to situate them in time. As he enthusiastically wrote in regard to this new historical discipline that represented, in his eyes, "scientific archaeology":

> One might call the new stage which archaeology has entered "the typological stage." Our next task is to establish *types*, to ascertain which of them are characteristic of each region, to search out a type's affinities,

> and to unfold its history. . . . Under the combined influence of two factors—practical needs and the craftsman's tastes—a great many forms appear, each of which has to struggle for its existence; if one of them succumbs, having failed to find what it needed to exist, another takes its place and produces a new series of forms. If any science at present needs its Darwin, it is comparative archaeology.[27]

Early on, then, the creation of forms had been viewed as resulting from the tension that existed between need and invention, between tradition and innovation. Hildebrand considered the transmission of characteristics to be a reproductive process, by which he meant a biological necessity: For inanimate objects and living species alike, only what would prove viable would be transmitted. In *The Prehistoric Peoples of Europe*, his major work, published between 1873 and 1880, and in which he became the first to attempt a synthesis of the knowledge recently amassed, he presented the biology of objects as the cornerstone of the emerging discipline of prehistorical archaeology.

> The stations during this development [of human civilization] are the types which correspond to the species in the organic world, though not the species as they are now, . . . but as they appear in paleontology, chronologically ordered. However, there is a difference in relation to the paleontological series in that in the culture-historical series one can more clearly distinguish the rise, the culmination, and the fall. From this, it is also clear that the types of cultural objects cannot be so sharply divided as the species in nature at the present day; one finds transitional forms in large numbers, but one gradually learns to distinguish certain forms which have become constant, while others show an uncertain fluctuation.[28]

Following Hildebrand, the Swedish archaeologist Oscar Montelius would take the connection between prehistoric typology and Darwinian paleontology to its furthermost point. Curator of the Museum of National Antiquities in Stockholm from 1868 to 1907, he attempted to reconstruct the chronology of European proto-historical times on the basis of the series of objects found in museum collections. In an article on the "methods and matter of prehistoric archaeology," he made an explicit connection between prehistoric archaeology—and specifically the typology of archaeological matter—and new developments in the natural sciences linked to the theory of evolution.

> The methodology of prehistoric archaeology has long been like that of natural science. Like the latter, the former has also now entered a new stage.
>
> The natural scientist is no longer content to describe the different species and to study their lives. He tries to find out the internal connection that binds them together and to show how one species has developed from another.
>
> What the species is to the natural scientist, the type is to the archaeologist.[29]

Like Hildebrand, Montelius was convinced that, as cultural creations, archaeological objects were comparable to living species viewed as biological creations. As in the natural sciences, archaeology identified and classified past species of objects and constructions that were, in Hildebrand's and Montelius' vocabulary, archaeological types. In both prehistory and the natural sciences, the core issue was that of the history of these creations. It was no longer enough, said Montelius, to describe and classify these objects. We had to understand the structure of the morphological relations linking these "species" or "types" to each other and thereby to grasp the way their characteristics were transmitted over time. To Montelius' mind, prehistoric archaeology and the natural sciences were basically engaged in the same kind of work.[30]

> The prehistoric archaeologist no longer regards it as his only task to describe and compare the antiquities from different countries and to investigate life in these countries in bygone days. He now tries to trace the internal connection between the types and to show how one has developed from the other. We call this *typology*.[31]

The method specific to prehistoric archaeology is thus typology, by which we mean the study of the evolution of types as constituted by archaeological creations. Montelius was very clear: The point was not to establish a catalogue that would define the types of objects or structures present in given regions at various times. The goal of prehistorical archaeology lay beyond such preoccupations; it was more than simply a matter of describing and comparing. Typology was the methodological tool with which to construct the biology of forms that it was for archaeology to establish. Archaeology lay beyond the field of history. It did not seek, according to Montelius, to learn exactly what had happened during various periods of the past, but to determine how the system of past material creations was structured, and how it had been transformed and transmitted over time. Here again, it was not enough to describe and compare. If past material creations evolved, then there was clearly something generating their evolution.

> That, as regards the productions of nature, it is possible to follow the evolution of one form or one species from the other has, of course, as we are all aware, long been known. But it is only recently that we have discovered . . . that a quite similar development can actually also be shown as regards the productions of human labour. This should interest the natural scientist so much the more as man is, of course, in himself, regarded as a production of nature, also an object of his studies.
>
> In actual fact, it is also an extremely wonderful thing that man should in his work be subject to an evolution governed by [natural] laws. Is human freedom so restricted that we could not freely make whatever forms we wished? Are we compelled to go, step by step, from one form to another that is only slightly different?
>
> Before one has investigated the matter more closely, one would certainly be tempted to reply "No" to these questions. But when one has become, by a thorough study, familiar with the remarkable phenomena of which I have just spoken, one finds that the answer must be "Yes." Evolution may proceed more quickly or more slowly, but man is always compelled, in his creation of new forms, to follow the law of evolution, as it applies to the rest of nature.[32]

The Failed Attempts at "Darwinian" Archaeology

In asserting that the typological evolution of artifacts was the result of natural laws identical to those of the biological evolution of living species, Oscar Montelius explicitly laid the groundwork for a biology of archaeological creations, but he failed to tell us what those laws were. As opposed to Hildebrand, he did not invoke the theory of natural selection to explain the evolution of archaeological types, effectively stripping Darwinian theory of its basic dynamism and reducing it to an energy-deprived evolutionism. In truth, it seems that no one at this time—we are in the last quarter of the 19th century—had any idea what to look for. The staunchest defenders of the ties between prehistoric typology and evolutionism could not envision the possibility that inanimate objects actually "struggle for survival." They turned to abstract representations of the societies that were responsible for their creation and fell back into a historicist kind of approach.

In England, General Augustus Pitt Rivers, who had avidly read *The Origin of the Species* when it was first published in 1859, promoted the idea of applying Darwinism to archaeology and ethnology. Toward the end of the 1860s, and above all in the 1870s, he developed his own concept of "cultural evolution." It was a very "Tylorist" approach that sought to reveal filiations in types found in existing prehistoric populations, the first step of which was to collect objects from these present-day "savages." In bringing together past and present "prehistoric" objects, Pitt Rivers was effectively trying to reconstruct a genealogy of forms that reached across both time and space. It was, he said, because formal connections obeyed the laws of evolution that they showed movement from simple to complex, and from homogeneous to heterogeneous. Pitt Rivers' perspective was not that far removed from Leroi-Gourhan's. He looked beyond the objects themselves in an attempt to arrive at the representations, what he called the "human ideas" underlying past and present material creations. In his view, the generalization of these ideas was analogous to the process underlying the evolution of species, in that these ideas were transmitted through *filiation*. As had been the case for Montelius, Pitt Rivers succeeded in demonstrating parentage across time and space in technique and morphology, but failed to provide a convincing explanation for it.

In the early stages of his work, the Swedish ethnologist Hjalmar Stolpe developed an approach close to that of Pitt Rivers. Like Hans Hildebrand and Oscar Montelius, he was part of the first generation of scholars who owed their scientific careers to the coming of age of the Stockholm museums' collections, in the 1860s. On the basis of the styles of "decorative" motifs found on ethnological objects, Stolpe sought to determine how the "formal connections," as Pitt Rivers called them, were established. It was his view that these connections were far more than the result of mere morphological proximity or parentage.[33] Stolpe wanted to bring out the basic laws underlying the evolution of stylistic representations. Like Warburg, he saw them functioning as a kind of pictorial language.

Stolpe's work is of interest in that it showed that the stylistic changes that took place over time obeyed a geometric sort of logic. Certain "background" elements gradually took on their own identity and became individual motifs that in turn divided and spawned variations, creating other kinds of figuration. Far from a trajectory moving from elementary to elaborate, the evolutionary dynamics of "ornamental" styles followed a complex, internal process involving the contradictory, complementary movements of separation and regrouping. Actually, Stolpe said, it was the same thing that was being perpetuated when representations were reproduced, but each time in a different form. Structurally, a form's message always included the same elements, but formally these elements were depicted in a specific way, for each was unique. Consequently, what a form signified was hidden, as it were, in the signifier, expressed as what Stolpe called a "crypto-glyph." "Ornamental" motifs were reproduced, he argued, to preserve not a particular form, but a particular meaning. And it was precisely because the point was to preserve meaning rather than a representation that iconographies were not only perpetuated over time, but that they maintained their "stylistic" coherence.

Henri Hubert's undertakings at the Museum of National Antiquities followed the approach initiated by Stolpe. It wasn't until the beginning of the 1980s that sociologists saw in Hubert—Marcel Mauss' soul-mate (Mauss coined the expression "work-twin")—a precursor of the sociology of time and raised him from the oblivion into which he had fallen.[34] Among archaeologists, he is primarily known for his two posthumously published works, one on the Celts, the other on the Germanic peoples. The material that he collected for these works prior to World War I constitutes an aborted attempt at an anthropological synthesis of historical and archaeological data.[35] In truth, Hubert was a man who undertook just about everything, and completed just about nothing. As the assistant curator of the Museum of National Antiquities, working under Salomon Reinach from 1903 on, his time was given to an "immense undertaking":[36] organizing the Saint-Germain museum's collection, which quadrupled between the end of the 19th century and 1920. During a period of almost thirty years, he committed his energies to acquiring new reference collections—especially in ethnology and non-European archaeology—and curating exhibits. His major contribution was the "Comparative Room" which, in its approach, was a forerunner to the presentation of the Moreau collection. In both instances, the idea was to display objects and series of objects in such a way as to show their formal relationships, and how types and styles were transmitted across time and space. As Hubert noted, the project aimed at turning the organization of the museum's holdings into a "microcosm" of great stretches of archaeological time.

Frédéric Moreau was a former overseer of the Bank of France who, in his retirement, began a career as an archaeologist at seventy. He amassed an impressive collection of prehistoric, proto-historic, Gallo-Roman and Merovingian artifacts from digs he conducted on his estate near the city of Soissons. The records of the finds made on these excavations fill the eighteen volumes known as the "Caranda Collection." By the early 1900s, Hubert saw in Moreau's compendium the material he needed to demonstrate the concept that would later serve as the basis for the "Comparative Room" display. The idea was to offer visual evidence of something that was not normally apparent but that was manifest in the morphology of objects. The compressed time of a museum exhibit would reveal not just the typological evolution of forms, but filiations and commonalities. This was all still in an unformulated, project stage of the biology of forms, but Hubert understood what exhibiting the processes of stylistic evolution meant for the conventional, sequential time of archaeology. In his draft of an introduction to the Moreau collection catalogue he wrote this:

> It moreover seems that this kind of collection can teach us more about the transitions from one period to another, the transformations, the mixing of customs and artifacts and the uncertain phases of metamorphoses than about eras in which civilizations appear to us to be pure, complete, and wholly unique.[37]

Notes and drawings held in the archives of the National Museum of Archaeology reveal Hubert's attempts to sketch out this project whose inordinate magnitude was, in truth, beyond his limits. In effect, it was not enough to set pieces side by side in glass cases, or to display them within vast archaeological periods. For the thousands of objects that were to be dealt with individually, there first had to be an interpretive framework that could bring out formal connections and, above all, the relational systems by which style could be linked to technique and ornamentation to function. And Hubert did try to draw up descriptive patterns that would reveal not only typologies or morphological classes, but also, and above all, connections within series, which is to say, degrees of commonality.

"It seemed to me necessary," he wrote, "within each class (of objects), to number separately distinguishable elements of form and decoration whenever no constant relation could be established between them. This distinction . . . was intended to bring out more clearly the *lineage* between interchangeable or associated patterns, the *relationships* between ornamentation and form, and the *borrowings* between them."[38]

As a matter of fact, the Saint-Germain archives hold several of Hubert's preliminary attempts at an interpretative framework for archaeological forms. He endeavored to match criteria used to identify decorative motifs (coded with Arabic numerals) with technical or morphological traits (coded with Roman numerals). His grids very much resemble the "pathos formula" tables Aby Warburg drew up between 1909 and 1911.[39] Like Hubert, Warburg tried, in his *Schemata Pathosformeln*, to match the "mimetic degrees" of forms represented in dancers, runners, etc., with representational types. It is striking that, in both cases, the grids remained empty. Archaeologist and art anthropologist alike proved incapable of filling them in.

Why did they come up empty? Why were these attempts to map out a biology of forms unsuccessful? Part of the response lies in Warburg's madness. It was directly from this failed attempt at a typology of "pathos formulas" that Warburg drew the material for his Mnemosyne Atlas, the "schizoid" collage of images gleaned from different times and different cultural contexts. What refused to fit into Warburg's and Hubert's grids was the conventional time that Walter Benjamin characterized as *historicist*. For forms, it is effectively impossible to reduce the clearly manifest structures of commonality or filiation to one single framework. It would require dozens, even hundreds of them for the same objects, because even though patterns and shapes can be grouped by types associated with specific times and places, what unites them does not occur in a time or a place, or if you prefer, the relationships among them all take place at the same time. To be sure, every time a shape is created—that is, physically, every time one is actually made—one effectively adds to a linear succession of objects, but the traits that are transmitted by or introduced into this shape have not traveled the same road; they have not directly proceeded from each other. The creation of an object is always a reinvention, one that might draw upon the most archaic way of doing things even when it appears to be thoroughly modern. What is new—what is created here and now—lies outside of history, for it deals with memory. As Maurice Blanchot said:

> The new, because it cannot take its place in history, is also that which is most ancient: an unhistorical occurrence to which we are called upon to answer as if it were the impossible, the invisible—that which has always long since disappeared beneath ruins.[40]

As Hubert clearly saw, everything that takes place happens in moments of transition, which are basically moments of recreation. But how can we depict what is going on in these "metamorphoses"? No one knows. The tools used in traditional history are of no help. As Walter Benjamin wrote in the sixth thesis of his "On the Concept of History": "Articulating the past historically does not mean recognizing it 'the way it really was.' It means appropriating a memory, as it flashes up in a moment of danger."[41]

The Cycles of Forms

Archaeologists have at their disposal an extraordinary mathematical tool with which to determine the degree to which the combinations of typological or stylistic traits found in artifacts resemble each other: seriation. When applied to extensive series of objects or structures, the method allows us to observe how archaeological types evolve over time. On a larger scale, it allows us to follow the changes in combinations of types that identify distinct periods in sequence. And yet, this method often encounters a major obstacle: We do not know, or know only very vaguely, the specific dates when types originated. Because they have nothing better to work with, most researchers simply assume that the order in which types follow each other, as determined by these methods of sequencing, corresponds more or less to their actual chronological order.

Bruno Wirtz, a mathematician friend of mine, and I undertook an unusual experiment of the "Millie's Camp" kind. We subjected a series of objects to a variety of sequencing calculations, with the idea of comparing the results with the actual fabrication dates, which we could pinpoint to the year. We used miners' safety lamps that had been manufactured between 1840 and 1975.[42] The lamps were an essential piece of equipment during the Industrial Revolution. The oil-burning technique that had been invented by the great British chemist Humphrey Davy in the early 19th century isolated the lamp's flame from the outside air so as to prevent gas explosions, which were frequent and often catastrophic in the mines of the early industrial age. Because the lamps were a crucial part of mine safety, their manufacture was placed from the outset under strict technical control, which effectively limited the number of parts. Throughout their history, miners' lamps were built along the same lines, with an oil vessel, a wick, and a light reflector. We thought that this would limit the number of variations, which in turn would make the changes that occurred over time stand out all the more clearly.

Our calculations produced results that were, to say the least, surprising. It became immediately apparent that a fundamental disconnect existed between the rate at which changes were introduced in these safety lamps and "real" chronological time. After a tentative first few years, the lamps produced in the 1850s incorporated bold innovative techniques that one would not have expected to see before 1880 at the earliest, if advances had been made at a steady rate from the start. This first burst of innovation gave rise to a wide variety of lamps, but by the 1880s they became increasingly standardized. The rhythm of change decreased, with lamps manufactured in the final twenty years of the 19th century evolving more slowly than "real" time, at a rate that was much closer, in fact, to what had been seen in the 1860s and 1870s. With the advent of newly developed techniques for burning oil, the first decade of the 20th century witnessed another spurt in innovative techniques, but it proved relatively short-lived. From 1910 on, the trajectory of typological time fell below that of real time and eventually came to a halt. Up to the very last lamps produced in the 1960s, especially in Eastern Europe, the models were still quite similar to those that had been manufactured at the turn of the century. In effect, the irregularity of typological time made it impossible for us to place miners' lamps in the actual chronological order in which they came into existence. Calculations made on the basis of stylistic proximity tended to group them by degree of formal resemblance, which in this case meant attributing types from different times to some "period," which had actually never existed as such.

We then turned our attention to the trajectory of technical characteristics. Once again, we compared the rhythm at which they evolved to that of real chronological time. And here, too, the results demonstrated that technical features developed independently of the times at which they were introduced. Some appeared like comets blazing across the sky, only to fade away almost immediately. Others settled in more slowly, and dwindled away more slowly. Still others emerged, then began dimming as if they were going to fade out entirely, only to come back and develop further before disappearing definitively. These various trajectories were observable at every moment of real time. At each and every point, their diversity produced a variability that our statistical calculations could not account for. As had been the case for types, it was impossible for us to place the technical characteristics in the chronological order in which they had actually been introduced.

The miners' lamps study demonstrated that, fundamentally, it is not possible for us to ascertain simultaneously an object's place in "real" time (which we'll call its date) and its place in typological or "archaeological" time (which we'll call its dating). These are in fact two distinct aspects that are not coincident. As Henri Bergson pointed out, the conventional way in which we represent history ignores the duality of time, which is as much duration, or accumulation, as it is transformation or change.[43] One could also say, along the same Bergsonian lines, that, at any given moment, time points simultaneously in two directions: toward the future, through the continual emergence of what is new and which alters what already exists, and toward the past, through the continuous accumulation of what is "already there," which increases the mass of what is old. Our conventional perception of time—that holds that each era has its own specific time, unlike any other—does not take into account this other aspect of time constituted by duration.

Thus, it is as if at any moment in time there existed, within a given group of types, a wide range of characteristics allowing for an immense diversity of subsequent typological trajectories; that is, of possible futures, or histories. Not until tendencies begin emerging within this diversity are we able to discern retrospectively evolutionary paths as such. Much as with gully erosion, whose ramifications spread further and further back as the flow increases, the more distinguishable these evolutionary paths become the more they create paths that reach backward in the development of typological attributes. This kind of typological evolution produces a wake, or more precisely, a drainage effect in its immediate past.

This strange phenomenon was very clearly described by J. L. Borges in a short piece entitled "Kafka's Precursors." Borges imagined himself combing through all of world literature, from as far back as ancient Greece or China, in search of passages that prefigured Franz Kafka's unique style. Such passages do exist. They allowed Borges to establish a literary filiation with Kafka that is hard to dismiss as serendipitous, even if it is complete fabrication. Borges concluded the undertaking thusly:

> If I am not mistaken, the heterogeneous pieces I have enumerated resemble Kafka; if I am not mistaken, not all of them resemble each other. This second fact is the more significant. In each of these texts we find Kafka's idiosyncrasy to a greater or lesser degree, but if Kafka had never written a line, we would not perceive this quality; in other words, it would not exist . . . . The fact is that every writer creates his own precursors. His work modifies our conception of the past, as it will modify the future.[44]

Tendencies of this kind have long been noted, especially in art history. In *The Life of Forms in Art*, Henri Focillon noted that styles have their own developmental logic that is not that of universal historical time. All artistic movements, he said, develop an internal periodization that usually involves moving from an initial, archaic, "experimental" age to a classical, "normalizing" age, and finally to a baroque period. None of these periods is truly homogeneous, for each is subject to varying forces. Thus, in every artistic period, the evolution of works of art reveals, "juxtaposed within the very same moment, survivals and anticipations, and slow, outmoded forms that are the contemporaries of bold and rapid forms."[45] Because of the disconnect between a "work's period" and historical time, "contemporary art works often belong to different 'ages.'"[46] In *History: The Last Things Before the Last*, Siegfried Kracauer perceptively pointed out that historians could not be credited with fathoming this temporal oddity in material works from the past, even though they dealt with them directly. This was rather the doing of art historians and anthropologists who "had become aware of the problematic nature of chronological time."[47] As he noted:

> Art works, or more frequently their elements, . . . can be arranged in the form of sequences, each composed of phenomena which hang together inasmuch as they represent successive "solutions" of problems originating with some need and touching off the whole series. One after another, these interlinked solutions bring out the various aspects of the initial problems and the possibilities inherent in them. So it would seem evident that the date of a specific art object is less important for its interpretation than its "age," meaning its position in the sequence to which it belongs.[48]

Thus, every typological evolution has its own internal logic on the basis of which stylistic creations build their own histories. In so far as each history is manifest in works peculiar to a given "style," it is unique, but all go through analogous stages that are more a matter of "situation" than of "period." As Kracauer noted, the sequence that moves from archaic, to classical and then baroque goes well beyond that of a simple chain linking periods of formation, expansion, and degeneration, which is how they are usually presented. Archaic stages explore the new potential for form generated by a stylistic system in crisis that has exhausted its creative resources. Classical stages present and develop these innovations by exploring combinations of possibilities; they also establish a general stylistic model that promotes uniformity. The arrival of the baroque stage is symptomatic of the narrowing of formal possibilities; because it is no longer possible to open new paths, all that is left is to enhance existing forms that have already been elaborated upon by recomposing them. There thus lies at the heart of this cyclical, typological time a process in which the creation of forms negotiates with what already exists. As Stolpe noted, the "connections between forms" are not derived from simple morphological proximity, but from a dialogue which results in each new form, each individually created piece, finding its place. There are two opposing tendencies in this dialogue: one directed toward simplification and standardization, the other toward diversification and deconstruction; one composes, the other decomposes. Because forms have to be constantly recreated each time that they are produced, because they always have to be renegotiated, each develops a history that is specific to it. This is the history archaeological creations generate themselves, through memory.

Forms and Memory

The American anthropologist and art historian George Kubler carried Focillon's observations further in a short book aimed at "the discovery of the manifold shapes of time": *The Shape of Time.*[49] When one examines the artistic works of any given moment in time, Kubler said, "the cross-section of the instant . . . resembles a mosaic of pieces in different developmental stages, and of different ages, rather than a radial design conferring its meaning upon all the pieces."[50] Kracauer, who had read Kubler, drew the obvious conclusions for history: "since simultaneous events are more often than not intrinsically asynchronous, it makes no sense indeed to conceive of the historical process as a homogeneous flow. The image of that flow only veils the divergent times in which substantial sequences of historical events materialize . . . . The upshot, is that the period, so to speak, disintegrates before our eyes. From a meaningful spatio-temporal unit it turns into a kind of meeting place for chance encounters, something like the waiting room of a railway station."[51]

We thus totally mislead ourselves when we imagine that the process of stylistic evolution rests upon chronologically continuous phenomena. The developmental effects of chronological continuity that we observe in any given typological sequence are but one effect of a more complex process that actually evolves at every moment of a form's development; that is, every time something is created. It is effectively at each of these moments that some new creation's tendencies are determined, be it an object or an immobile structure: It can break with what precedes it or, conversely, carry it forward; it can move down the path of innovation or can solidify or even permanently establish inherited characteristics. The creation of every new object sanctions the choice made between these various options, and it is the chain of orientations that sustains typological evolution. The cumulative process yields a stylistic capital that can be drawn upon for the creation of new objects. Stylistic time is fundamentally a matter of transmission.

Thus, at work in these "self-determining" phenomena that are the processes of typological evolution is a kind of intrinsic memory, or memorization process, for the past has to have some sort of identity if it is to be drawn upon for the creation of something new. And there emerges from this return to the past not only the identity of present creations, but also what are commonly referred to as typological lineages. It behooves us now to turn our attention to this essential phenomenon. We are effectively dealing with how memory functions, and once again it matters little whether it is the memory of the mind or the memory held by archaeological matter. What is most important is the manner in which the past—that which exists—negotiates with the present, with that which is new.

How does that work? On the basis of a process he called *repression*—by which the present buries past events in the unconscious[52]—Freud described the phenomenon in a letter to Pastor Pfister: "All repressions are of *memories*, not of experiences; . . . when one is theorizing one should always try to find out what lies behind the complex, not make a frontal attack."[53] We need to reflect a moment on this formulation. There follow from it two important implications with regard to the process for evaluating the past that is at work in memory. First, as Freud noted, the point at which an event takes place is not necessarily the point at which it acquires a particular meaning. This is a phenomenon that is common to the processes of historical memory. We know that a brief or lengthy period of time can go by between the moment when an event takes place and the moment when it is recognized to be significant within a history that materializes subsequent to it. Archaeologists, too, know that the time when an archaeological object is created is not the time when its function is decided upon. That is determined by use, as witnessed by the innumerable examples of archaeological finds that have been reused, recycled, or reoccupied. There exists, moreover, a wholly comparable phenomenon in the processes of typological or stylistic evolution. In "classical" stages, forms and motifs of "archaic" phases become established in a generic way, not simply because they are transformed, but rather, as Stolpe pointed out, because they are "reread" or "renegotiated" when they are reproduced or redesigned in the styles of subsequent phases.

The second implication of Freud's observation is more surprising for our study. It holds that the moment when a fact is recognized as having a particular meaning is not experienced as a moment in the present, rooted in its own time. Rather, it is defined, to use psychoanalyst André Green's terms, as "retrospection through identity and difference."[54] This is a crucial point: If there actually is some filiation between typological characteristics within a stylistic sequence or, to approach the issue from a historical point of view, if there does exist a sequence of events within a given period, the sequencing is not the result of some chain effect whereby each archaeological creation or each historical event would simply be the extension of those that immediately precede it. The spontaneous creation of this history is, in each case, the result of retrospective evaluation of the past, which, as opposed to what one might believe, occurs in a time that is not the same place in which the negotiating happens. Rather, it takes place in some timeless space that cannot be chronologically determined because, as the place where the identity of events or archaeological creations is negotiated, it lies between past and present; in other words, nowhere.

Thus, when a new model of a type of object was created in some historical periods, its creators were not, strictly speaking, seeking to design an object "from their own time." Fundamentally, they were still looking to reproduce the same "concept," perhaps by adapting the object to restrictions newly in place or to demands made by those for whom it was intended. In other words, that which exists—that is, the formal characteristics that have accumulated over time—is reconfigured, thrown back into play with every new creation. As a result, repetition and reproduction are present everywhere in processes of typological evolution, which is what allows for the great variety one finds within typological time that establishes the rhythm of techno-morphological change. It is because what already exists is renegotiated with each new creation—which is to say at each moment of typological time—that the resultant re-evaluation invariably yields a great diversity of responses ranging from simple reproduction to complete transformation. In this context, it is now easier to understand why Freud speaks of "deferred structuring": As is the case for events of the past, the significance of archaeological constructs is not established when they appear, but throughout their various re-evaluations over the course of time. It is the future that gives meaning to the past, not the past itself.

Thus, from the moment the process of typological formation begins, the forms produced are defined in relation to what has been inherited from a past—the stylistic or typological capital—that is not simply reproduced, but retrospectively re-evaluated in terms of difference and identity. What happens then is fascinating, because the recomposition of forms that characterize the process of stylistic or typological creation is fundamentally unstable, in that it is caught in a particular tension field. At stake with every new creation is the very existence of the past as a viable entity that has come down to the present: The development of new forms must not do away with the past, either by eliminating existing forms (that is, by totally abandoning them in favor of something entirely different) or by absorbing them (that is, reproducing them almost exactly as they were). Each time a craftsman creates a new form (a new pot, or tool, or house), something both identical and different adds to the number of material creations in use and gives them new life, for objects are fragile and consequently mortal. Objects create their own history, their own typological trajectory, out of the constant reproduction of material creations that waver between diversity and homogeneity. It is then easy enough to understand why forms die out once they are stabilized or become set.

It follows that, at any given point, no moment of typological time is heading anywhere in particular, as we can see from the typological evolution of objects that archaeology allows us to reconstruct. Each moment in time is more an expression of potential than a combination of specific typological attributes. This explains why we can clearly identify, most notably when new forms or new types of objects appear on the scene, the features that are going to combine and become generalized in subsequent phases of their development. But just as these particular features are at a germinal stage, all the others are as well. There is no telling, at any given time, which of them will be developed and which of them will be abandoned. Only their future typological evolution can tell us that, and that future will be one which, at any point, could have been totally different from the one it actually became.

The future of typological attributes is at once predetermined—because it is transmitted by the past—and undetermined, in so far as the identity it acquires depends on what happens subsequently.[55] Once in place over a period of time, once it has been reproduced and transmitted as new objects are created, its past identity takes on substance, so to speak, retrospectively. As new types of objects are being created, what was originally just one of the possible technical solutions or stylistic innovations becomes, in the course of a few decades, the standard form that corresponds to the "concept" that had been sought. What had been just one attempt among many becomes a kind of reference point that will be extremely difficult to change thereafter. The process leads directly to the baroque, to the over-development of morphological detail at the expense of overall form, which ceases to evolve. It is not so much a matter of having exhausted the resources of an existing stylistic register as of the impossibility of altering it.

Thus, throughout the negotiating process involving innovations introduced through new material creations, there exist different periods or eras that depend on the conditions under which the re-evaluation of the past in the present occurs. This explains why the baroque always comes at the end, when the stylistic or typological stock has begun to rigidify. It also explains why there are generally no transitions from one style to another, in so far as the baroque—which is a form of over-accumulation—is an indication that a given stylistic or typological system has entered a mode that has "reached its limits," in which it is no longer possible for it to be reformed anew. This is the point at which it appears to be invaded by parasitic forms and features that seem to have come out of nowhere: The movement from one style to another always seems incoherent, to have been "intruded upon."

We now see as well why "archaic" stages are characterized by a broad range of types and features, whereas classical phases in all cultures seem so "standardized," so "tame." Here again, the key to this phenomenon lies in the process of negotiating with the past that has taken place in the present, which occurs at every instant of typological time; that is, with the creation of each new object. Objects are quite simply the reification of this dynamic, its inscription within matter. This explains why the typological trajectory of objects is so irregular: It is marked by a succession of objects that are all different from each other, because they are individually unique. As Freud said, an individual life is composed of a series of "innumerable I's," each of them distinct.

We need to get back to the time of forms, which, clearly, is difficult to get a handle on. Strictly speaking, the relation between the past and the present does not take place in the present. Past and present actually coincide and illuminate each other in a reciprocal "time outside of time." When added on to the past, the present produces effects of chronological continuity or historical trajectory—which in reality does not exist as such—in a time that is essentially not bound by time.

This is very much the reason why the time of forms is not linear, why it is not made of a direct, chronological succession of morphological innovations. For example, when the urban fabric of an ancient, then a medieval, town was remodeled in the modern, or the contemporary, era, or when the parceling of a rural landscape was reconfigured, past and present confronted each other in a space defined by a dynamic of function or identity that was not linked to any particular moment in time, for the dynamic embraces all of time. When new houses have to be built or older ones renovated in a town's center, it is above all because that is still where people want to live. Similarly, if agricultural lands are to be reallotted, it is because they are still to be used for growing crops . . . . This contradictory relationship generates history. In other words, the tension makes it clear that an identity has been carried forward; and, obviously, it can emerge only after the fact because, paradoxically, it is the generally unexpected association of elements of these occupied spaces that brings out the continuity. History in progress is invisible to us. As historical phenomenon, and also as knowledge of the past, history takes shape in the joining of, in the bringing into relation of events that are fundamentally discontinuous. Absence and oblivion create the hiatus that is necessary for them to be joined, the blank space that gives meaning.

Notes

1. An entry in his *Ricordi*, dated December 8, 1895, while he was exploring the ruins of Indian troglodytes in Mancos Colorado, reads: "Pneumonia" (my obsession)." Aby Warburg, *Le Rituel du Serpent. Art et anthropologie. Récit d'un voyage en pays pueblo* (Paris: Macula, 2003), p. 137.

2. Aby Warburg's unpublished notes for a lecture given in Kreuzlingen, in 1923, on the "snake dance." Quoted in Philippe-Alain Michaud, *Aby Warburg and the Image in Motion,* trans. Sophie Hawkes (New York: Zone Books, 2004), p. 301.

3. Georges Didi-Huberman, *L'Image survivante. Histoire de l'art et temps des fantômes selon Aby Warburg* (Paris: Minuit, 2002), pp. 134–6 and drawing 13.

4. Quoted in Philippe-Alain Michaud, *Aby Warburg and the Image in Motion*, trans. Sophie Hawkes (New York: Zone Books, 2004), p. 332.

5. Ludwig Binswanger and Aby Warburg, *La Guérison infinie. Histoire clinique d'Aby Warburg*, trans. Maël Renouard and Martin Rueff (Paris: Payot et Rivages, 2007).

6. Warburg, *Le Rituel*, p. 60.

7. In his unpublished note for his lecture in Keuzlingen on the "snake dance" in 1923, Warburg wrote that, for him, "image and words" ought to help people

"defend themselves from the tragic aspect of the tension/split between [word added: magical] instinct and inhibition/discursive logic." Quoted in Michaud, *Aby Warburg*, p. 296.

8. Quoted in Michaud, *Aby Warburg*, p. 296.

9. Quoted in Michaud, *Aby Warburg*, p. 301.

10. *Kreuzlingen Diary*, reproduced in Didi-Huberman, *L'Image*, picture 79–80.

11. In Michaud, *Aby Warburg*, p. 294.

12. Ernst Hans Gombrich, *Aby Warburg. An Intellectual Biography*. (London: University of London, Warburg Institute, 1970).

13. Aby Warburg, *Schlangenritual. Ein Reisebericht* (Berlin: Klaus Wagenbach, 1988).

14. Michaud, *Aby Warburg*; Warburg, *Le Rituel*.

15. Near the end of his life, Warburg summarized his approach thus: "Sometimes it looks to me as if, in my role as psycho-historian, I tried to diagnose the *schizophrenia* of Western civilization from its images in *an autobiographical reflex*. The ecstatic Nympha (manaic) on the one side, and the mourning river-god (depressive) on the other." *Diary*, April 3, 1929, quoted in Gombrich, *Aby Warburg*, p. 303.

16. Quoted in Michaud, *Aby Warburg*, p. 301.

17. On the title page of the manuscript on the "snake ritual," Warburg wrote: "It's a lesson from an old book: the kinship of Athens and Oraibi." Quoted in Michaud, *Aby Warburg*, p. 295.

18. Johann Joachim Winckelmann, *Geschichte der Kunst des Altherthums* (Dresden: Walther, 1764).

19. Quoted in Michaud, *Aby Warburg*, p. 302.

20. Carlo Severi, "Warburg anthropologue, ou le déchiffrement d'une utopie. De la biologie des images à l'anthropologie de la mémoire," in *L'Homme, revue française d'anthropologie* (2003), pp. 78–85.

21. Edward Burnett Tylor, *Primitive Culture* (London: John Murray, 1871).

22. Warburg, *Le Rituel*, p. 60.

23. Quoted in Michaud, *Aby Warburg*, p. 304.

24. Warburg, *Le Rituel*, pp. 131–3.

25. Robert Klein, *La Forme et l'intelligible (*Paris: Gallimard, 1970), p. 224.

26. Giorgio Agamben, *Image et mémoire. Écrits sur l'image, la danse et le cinema*. (Paris: Desclée de Brouwer, 2004), p. 9.

27. Hans Hildebrand, *Den vetenskapliga fornsforkningen, hennes uppgift, behof och rätt* (Stockholm, 1873), quoted in Bo Gräslund, *The Birth of Prehistoric Chronology. Dating Methods and Dating Systems in Nineteenth Century Scandinavian Archaeology* (Cambridge: Cambridge University Press, 1987), p. 101.

28. Hans Hildebrand, *De förhistoriska folken i Europa. En handbok i jämförande fornkunskap* (Stockholm, 1873–1880) quoted in Gräslund, *The Birth*, p. 102.

29. Oscar Montelius, "Den Förhistoriske fornforskarens metod och material," in Gräslund, *The Birth*, p. 102.

30. In the same article, Montelius stated that prehistorical archaeology involved "a method which, in actual fact, is nothing else than that of the natural scientist, although it is applied not to the productions of nature but to relics of man's prehistory." Montelius, "Den Förhistoriske," 27, in Gräslund *The Birth*, p. 103.

31. Oscar Montelius, "Typologien eller utvecklingsläran tillämpad pa det menskliga arbetet," in *Svenska Fornminnföreningens Tidskrift* in Gräslund *The Birth* (1987), p. 103.

32. In Gräslund, *The Birth*, p. 103.

33. Hjalmar Stolpe, *Collected Essays on Ornamental Art* (Stockholm: Aftonbladets tryckeri, 1927).

34. François-André Isambert, "Henri Hubert et la sociologie du temps," *Revue française de sociologie*, xx (1979).

35. Henri Hubert (Paris: Albin Michel, 1950); Henri Hubert, *Les Germains* (Paris: Albin Michel, 1952); Olivier, "Nouvelles recherches."

36. Mauss, "Avertissement" in Henri Hubert, *Les Celtes et l'expansion celtique depuis l'époque de la Tène et la civilisation celtique* (Paris: Albin Michel, 1932), p. 24.

37. Henri Hubert, "La Collection Moreau au Musée de Saint-Germain," *Revue archéologique*, XLI (1902).

38. Hubert "La Collection Moreau," p. 178.

39. Didi-Huberman, *L'Image*, drawing 47.

40. Maurice Blanchot, *The Writing of the Disaster*, trans. Ann Smock (Lincoln: University of Nebraska Press, 1986), p. 37.

41. Benjamin, *Selected Writings*, vol. 4, p. 391.

42. Laurent Olivier and Bruno Wirtz, "Recherches sur le temps archéologique: l'apport de l'archéologie du présent," in *Antiquités nationales*, 35 (2003).

43. Henri Bergson, *Matières et mémoire.* (Paris: Presses universitaires de France, 1896).

44. J. L. Borges, "Kafka and his Precursors," trans. James E. Irby, on-line: http://sillysoft.net/forums/viewtopic.php?t=1214

45. Henri Focillon, *The Life of Forms in Art* (New Haven, CT: Yale University Press, 1942), p. 65.

46. Focillon, *The Life of Forms*; Kracauer, *History*, p. 143.

47. Kracauer, *History*, p. 143.

48. Kracauer, *History*, p. 144.

49. George Kubler, *The Shape of Time* (New Haven, CT: Yale University Press, 1962), 11. "I owe my first concern with the problems set forth here to the works and person of the late A. L. Kroeber," Kubler said. Their correspondence began in 1938 (p. 121).

50. Kubler, *The Shape*, p. 25.

51. Kracauer, *History*, pp. 149–50.

52. Freud's formulation of this psychoanalytical concept is couched in archaeological terms: "Repression, which simultaneously renders psychic matter inaccessible and maintains it intact, can best be compared to the burying of ruins, such as was Pompeii's lot, until spadework unearthed it." Sigmund Freud, *Le Délire et les rêves dans la Gradiva de W. Jensen*, trans. Paule Arbex and Rose-Marie Zeitlin (Paris: Gallimard, 1986), p. 170.

53. Sigmund Freud, letter dated January 10, 1910, in *Psychoanalysis and Faith. The Letters of Sigmund Freud and Oskar Pfister*, Ed. Heinrich Meng and Ernst L. Freud, trans. Eric Mosbacher (New York: Basic Books, 1963).

54. André Green, *Diachrony in Psychoanalysis*, trans. Andrew Weller (London: Free Association Books, 2003), p. 16.

55. In regard to *origins*, Walter Benjamin very rightly stated that "they never make themselves known to us in their naked existence, clear and factual, and that their rhythm is only visible in the dual perspective of their pre- and post-history." He further noted that "although origins very much represent a historical category, they *have nothing to do with the genesis of things*. Origins do not refer to what things that already exist are becoming, but to the process of coming to be as things evolve and decline . . . . Origins demand to be recognized, on the one hand as *restoration* and restitution, on the other as something that is, by that very fact, always incomplete, *always open-ended*." (emphasis added)

CONCLUSION

What Remains

ROOTS . . . The first little silk hairs, maybe two or three thousand years ago . . .

A slow progression marked the thousand-year history of this tree's gestation, followed, seventeen or eighteen hundred years later, by major development.

As the branches and trunk grew, the roots kept penetrating farther down in their perpetual search for sustenance,

Filling the space left by the extracted matter.

From the constant, relentless, sharp little sounds of their expansion and the creeping, heaving movement of matter came the vibrations, the hum, and the crackling of growth.

IN THE VISCERA, the old misery went on rotting. Walls crumbled away, yielding the black milk necessary for the trunk to grow.

Then, from the vertical roots, there emerged in black and white the darker human mass of tuxedos made of foliage and fruit.

It was hot and humid; everything smelled of urine and sweat;

The images were blurry, flickering, and dusty;

The constant, secret creaking of the wood reverberated in the thick air and filled the roots, like blood being pumped into the womb.[1]

CONCLUSION

The Ledger of Time

Giuseppe Penone described it in his sculptor's way: History is alive because it is made of matter, and because matter that is alive grows. The past, "the old misery," because it has been crumbling away from the very beginning, is secretly nourishing the present with its "black milk." An irrepressible movement that comes from the depths of matter fills every nook and cranny. As it grows, the present is pervaded by the creaking of the past, enlarged by all that growth untiringly brings. And, as it expands, the present reaches farther and farther down in search of what it needs to grow. Everything that lives creates memory inscribed in matter that is constantly evolving, decomposing, and recomposing. Penone's obsession with ever-evolving vegetal matter and with the wood in which time is recorded in rings (also called growth rings) brings to mind one of Bergson's observations: "everywhere there is life, there is an open ledger on which time is inscribed."[2]

The ledger in question is matter. Artifacts are made of living matter. They are a proximate form of that true nature that the Japanese painter Hokusai sought throughout his life.[3] Archaeological remains are the very substance of history, a history manifest not in chronicled events but in ephemerality. "A blade of grass is as graceful as a tree or a mountain," said the Spanish painter Juan Miró, who tried to "explode" painting by reducing it to its essential forms. He went on to say: "Aside from primitives and the Japanese, everyone scoffs at these divine elements."[4] Life is miraculous because it is at once temporary and indestructible, both timeless and constantly new. Artifacts stand as proof of that. They are the place where life's work is inscribed and becomes legible, taking the imprint of what we call time. These are things of no especial value, the ordinary, relatively interchangeable scraps with which museums are, already, inundated. And that is precisely why they are unique and precious. As George Pérec said, they are our fortune.

> The time that goes by (my history) deposits residue that piles up: photographs, drawings, long dried-out felt-tip markers, shirts, deposit and non-deposit bottles, cigar wrapping, cans, erasers, post-cards, books, dust and knick-knacks, what I call my fortune.[5]

What is it exactly that remains? We have to go back to the beginning, to the moment when the things intended for archaeological study are unearthed, identified, and categorized: to the dig. At bottom, archaeology is a process of dissociation. The archaeological has to be separated from the non-archaeological, anthropogenic remains have to be extracted from the natural accumulations in which they lie. We labor under the belief that we are wresting man-made objects from the non-human matter into which they disappeared long ago, but even within the matter of artifacts, the distinction between human and non-human is not at all as clear as we tend to believe. In fact, the older material remains are, the more the separation between the two becomes unclear, the more the anthropogenic has been absorbed, as it were, into the natural. It is impossible to know where to draw the line. There is no clear surface that we can ascribe to a particular moment of the past. Quite the opposite is true: Everything in the earth is floating in uncertainty, in a realm of maybe. We dig by sight, into extraordinarily rich and complex matter where the past we are looking for is closely entwined with its pre- and post-history.

Consequently, every dig is a necessarily false proposition, for the act of extraction is an act of amputation, of simplistic elimination. It is obvious that what we cast aside is not at all dead matter, not some sterile accumulation that we just have to remove to bring back in full the ancient remains that lie buried there. In truth, what we are casting aside is the entire subsequent history of the one, hypothetical moment of the past that we are looking for: its memory, its uninterrupted retranscription, until the point of oblivion. Conversely, what we will be digging into, farther down, below it, is everything that prepared and pointed to its coming and that took on meaning because it did come. Digs escape our grasp because, from the outset, we have held to the illusion that since something from the past remains, it will be immediately intelligible to us. And that is where we have been misleading ourselves: This past that emerges for us, because it has been preserved, is withdrawn from us at that very same moment. For something tangible to come down to us, everything around it had to have disappeared without leaving a trace; for this something to be preserved, it had to have been physically changed. For us to be able to identify it, it had to have been absorbed into something foreign to it. We have no concepts with which to deal with this, other than that of a paradox.

History and Memory

Our practice of archaeology is still tied to the traditions of Antiquity. We have never been able to free ourselves from archaeology as cultural history, principally because one of the basic hypotheses of the discipline holds that the past that lies outside of recorded history is historically knowable. This well-entrenched tradition has it that artifacts themselves bear witness to the identity of the specific, historical moments they emerge from. This is what Hippias believed, convinced that information that came to us from the past was necessarily unambiguous. According to him, once located, disassembled remains of past material creations could be pieced together through stylistic similarity, and the characteristics of the civilizations that had produced them reconstructed. And, by aligning them by degree of stylistic proximity, we could also establish the order of the periods that had left their mark on the history of civilizations, from their rise to their decline.

This was effectively Winckelmann's approach.[6] Conventional archaeology posits that artifacts bear witness to cultural history. Reconstructing their order in time and space is the same as tying together the elements of a narrative that relate the story. The archaeologist's mission has not varied since its most distant beginnings in classical Antiquity. We are still recounting the history of men and their civilizations on the basis of the creations they left behind. Once again, this simply doesn't hold up, if only because the time of artifacts is not recorded in annals but in matter. This material memory does not hold moments, but durations. In these durations there lies an accumulation of everything that remains, things that date back to the most varied of origins and that constitute the fertile ground that holds the seeds of that which may someday come.

The second basic hypothesis of this approach posits that the past that lies outside of recorded history is naturally intelligible to us. Through some quality peculiar to humans, we are able, because we are human, to recognize and understand the thoughts, acts, and material creations of those who came before us and have since disappeared. It is a firmly rooted postulate as old as the discipline itself. Since first formulated in the 18th century, it has remained the cornerstone of prehistoric archaeology. And yet we can clearly see that the anthropogenic and the cultural are never absolutely separated from the non-anthropogenic and the natural.[7] On the contrary, as Bruno Latour noted, we are everywhere surrounded by hybrid structures and processes that belong to the natural order and the cultural order alike. This is clearly the case in archaeology. Hybrid entities are not simply compound objects or half-breeds; they are monsters. They are forms that we are loath to conceive, creatures that are beyond our control. To conjure them up is to set them free, after which it is nearly impossible to rid ourselves of them.

In a traditionally historical perspective, the incorporation of anthropogenic creations into natural matter, the hybridization of the cultural and the natural, the multitemporal memory of matter, are not just basic obstacles to our conventional apprehension of the past; they are also and above all unimaginable. Before the sciences—the so-called hard sciences—began taking an interest in the history of matter, in the memory held in nature, we were able to hold on to the notion that the history of mankind and the memory of his social groupings were wholly discrete. That notion is being challenged, and the obstacles we have run up against have told us something particular about the nature of archaeological remains, which are not what we would have them be. There may well be, in this questioning of the way social sciences operate, an opportunity for us to create a new basis for apprehending the past by eliminating the (hi)stories with which it is encumbered.

The social sciences now find themselves mired in a "crisis in time." Conventional, historical time is under fire. As early as the 1980s, the immobile time of anthropology and ethnology, which is devoid of history, was viewed as problematic.[8] Anthropology and ethnology had to come to terms with that. They were by definition ahistorical disciplines given to the study of the particular structure of human societies, not to changes in them.[9] The study of the history of cultures and civilizations fell to prehistory and archaeology, which sought to produce historical time for periods for which none was known; that is, to fill in the black holes of the past, before or beyond the existence of recorded history. Now we know that what plagues the social sciences is not some time in history that is missing, but history that is missing (i.e., devoid of) time.

As Walter Benjamin pointed out, conventional history, or more precisely put, the traditionally historicist perception of history, drains time of its substance. It is hardly surprising, then, that this "crisis in time" has been accompanied by another crisis that involves historical objects and thus directly affects archaeology. We no longer know how to deal with artifacts and ways of organizing them that once seemed so obvious to us; we no longer know what grid to place them in. We can sense only that they have slipped out of our grasp and that the way we have been dealing with them can no longer tell us what they are.[10]

Of course, we could just carry on as before, as if nothing had changed. We could even default to this retrograde approach and use the new finds we keep making to fill in the openings through which the past has been slipping away. We could take our cue from the people at Chernobyl and build some tightly sealed vault that, for as long as we could maintain it, would keep the irrepressibly spreading toxic substance of the past from seeping out. But it seems to me that we need rather to heed the argument in Walter Benjamin's "Theses on History": We have to abandon the historicist's "empty, homogeneous" time in favor of the full, heterogeneous time of nowness. Once we recognize that archaeological time is not the linear, sequential time of the historicist, all of archaeological time appears to us in the here-and-now time Benjamin referred to as *Jetztkeit*. In effect, archaeology is not interested in past events in themselves, but in the memory that is constructed over time through repetition and re-evaluation. In this respect, it is the basic mission of the discipline to define itself as an archaeology of the present.

The problem lies not with history, but with time. Lyell and Darwin, and then Freud, reversed the directional arrow. They showed that to reach the past, we have to start, not from the past, but from the present, and from the present alone. Information about the past is contained in the present, not in the past, which no longer exists and which is definitively inaccessible to us. There exists in this domain an odd genealogy of ideas that moves from Nietzsche to Warburg and Benjamin, and which, above all, links Darwin to Freud. It reveals another approach to time, a dizzying, dangerous approach. It is an understanding of time that situates its subject in the present, recognizing it to be the node where past, present, and the yet-to-come intertwine.[11] By dealing with the matter in which memory is contained not as witness to the past but as a sign of it, surviving traces and repetition become an object of study. There is no name for this approach. As Agamben said in speaking of Warburg's work, it is very much a "nameless science," one that involves fields that are seemingly foreign to each other, like psychoanalysis, paleontology, art history, and archeo-geography. It has no name because it has yet to be formulated. It is a discipline of the minute and the trivial, one that finds its matter in what history discards, in its rags and its trash. It exists in nowness, as it seeks out that which lies in the creases of the ephemeral, in that which does not last but persists. I have no idea what to name this approach, but I know the field it belongs to: The approach that deals with memory is that of archaeology.

Redirecting the study of the past toward memory awakens echoes of the cataclysms generated by the now distant revelations of Darwinism and psychoanalysis. Just as we have been forced to recognize that man does not control the way nature evolves (Darwin), and that "the 'I' is not master in his own house" (Freud), we must now recognize that we are not masters of our material creations, which live their own existences as things and develop their own memory. As Paul Ricœur pointed out, far from undermining efforts to understand the past, relocating the locus of the meaning of history actually revitalizes these efforts: "suggesting that there may be a paradoxical structure to history" only posits more acutely the questions of history's meaning and its intelligibility.[12]

For There to Be Memory, There has to Be Forgetting

A metaphor from Borges' *Ficciones* allows us to grasp why the movement of history is invisible to us and why, in spite of that, our very blindness provides us with an opportunity to give it meaning. In "Funès, the Memorius," a Uruguayan Indian who has fallen off a horse is struck with a strange infirmity: He now remembers everything he has seen in its minutest detail:

> A circumference on a blackboard, a rectangular triangle, a rhomb, a diamond, are forms which we can fully intuit; the same held true for Ireneo, for the tempestuous mane of a stallion, a herd of cattle in a pass, the ever-changing flame or the innumerable ash, the many faces of a dead man during the course of a protracted wake.[13]

Funès finds himself overwhelmed by these shapes that usually remain invisible, instantaneously forgotten. He can see them because his memory pathologically holds a clear picture of things and beings at every moment in time, whereas the vision we have of them is generally intermittent. Borges described his character thus: "He was spectator to a multiform, fleeting, excruciatingly precise world." This world of material reality is the one we would see if we were able to seize the structure of every instant in time. It is a world teeming with details that are constantly being reconfigured into other shapes, which then burst and form new shapes, like a kaleidoscope. The "intrusion-filled" trajectory of miners' lamps gives us a sense of this multiform universe in that, like Funès, we have a detailed record of each of them at specific times. Better stated, it is because we are able—through what are effectively unusual means—to situate each of these lamps in time, that we can glimpse a small part of what Funès saw.

Funès is overwhelmed by the overabundance of these images because, paradoxically, he has lost his ability to recognize reality. Recording reality in total effectively denies it its reality; a history that records absolutely everything that happened does not allow for history. We are able to create meaning between two moments precisely because there are gaps. Better stated, meaning emerges in the field of tension that gaps generate. This is the very principle of association developed by Freud. Borges' short story deals with what is actually a profound issue, that of temporal continuity. For there to be memory—that is, for a series of events to have meaning in time—they have to be discontinuous, unconnected. Were archaeology to be complete, should archaeologists come to identify every stage of every site of every period, they would suffer from the same affliction that struck Funès: They would be incapable of reconstructing history. All they could give us to see would be some ever-changing yet always similar multifaceted turbulence, the vibration of matter thrust into time. We need to realize that history, understood as a process that generates meaning from vestiges of the past, is not based on the reconstruction of a series of events or of archaeological facts over time. History is memory creation. What is to be deciphered is located between the fragments.

This is precisely the case for cinema, where images are put together to create a story. As Jean-Luc Godard recently said:

> Actually, it is not an image we see, but a coming together, or as Walter Benjamin said, something in the distance, however short that distance. In cinema, we use a technique known as shot—reverse shot, in which

> we show two faces one after the other. In truth, you are seeing the same face twice, because to exist, the reverse shot needs the story in order to bring in a third element that I would call the real image or the real story, which is the picture or the story.[14]

"Nature Is Wont To Hide"

We have to understand that artifacts cannot be understood as witnesses to the past we seek. "Nature is wont to hide," said Heraclitus, adding, just as mysteriously: "If you do not expect the unexpected you will not find it, for it is not to be reached by searching for it or following a trail."[15] Indeed, the past lies hidden in archaeological remains that, because they are hidden, are not necessarily telling us what their visible form might suggest. We have to remain constantly on the alert for the unexpected, which is effectively difficult to find because it lies hidden beyond its thoroughly banal appearance. As in Edgar Allen Poe's "The Purloined Letter," it is hidden even though it is in plain sight. As we learned from what happened to the Auschwitz scrolls, to hide things, to bury them, is basically to leave traces behind. It behooves us to search untiringly for these signs of what lies hidden, which is the very stuff of archaeology. We have to seek out not only artifacts that lie buried in the ground, but what is inscribed within them.

Traces and signs; medical science tends rather to speak of symptoms. For a physician, a symptom is not merely a signal, it is above all the visible manifestation of something that is not visible, such as an illness or a pathology. In Greek, "symptom" means "that which befalls together." Its etymology reminds us of the connection between the pathology everyone similarly affected shares, and the distinct ways in which that pathology is manifest in given individuals. This should help us understand that artifacts are not witnesses to past history, but rather symptoms indicative of the living memory of the past. Experience has taught us that symptoms are meaningful, even if the person who has them is unaware of it. They indicate the presence of something that basically is beyond his control, because even though it is his, he is being subjected to it. Thus, while there exists for every pathology a general typology of symptoms, individuals react to illnesses differently. In some form or another, the pathology that inhabits an individual becomes distinctly his, part of his personal history. A symptom is thus more than a sign or a trace; it is a reaction to something gone awry or, better put, the compromise between two fundamentally irreconcilable situations, the invading illness on one hand and the absence of illness on the other. It is the arrangement that has been reached to allow the individual to carry on in spite of the irrepressible illness present within him.

When we say that archaeological creations function like symptoms, we mean that they are the signs of a memory that is still being shaped in the here-and-now, which is why these creations always appear highly structured. Their structure reflects a lot more than just their identity as archaeological manifestation or cultural creation. In this perspective, archaeology needs to be redefined as a discipline specifically attuned to the symptoms of material memory. It is hardly gratuitous to point out, with regard to the memory of the psyche, that it was through the study of the symptoms of memory—in particular, as revealed in dreams—that Freud was able to develop his approach to psychoanalysis. It was after he had grasped the way in which the symptoms of memory functioned that he was able to deduce his theory of the unconscious. An archaeologist's mission is quite similar: Beyond the artifacts themselves, the object of archaeology is time's unconscious, for it is precisely because there is such an unconscious that forms from the past survive and come down to us.

Understanding how buried or hidden memory is constructed essentially involves grasping the mechanism of repetition, or reiteration. Each time a shape is created from matter, it is in fact reproduced. At any given moment, forms are simultaneously being repeated, perhaps millions of times. The repetition itself is transformative in that the creative act is basically one of negotiation: What emerges must find its proper place among all the previously created things that surround it. As Warburg pointed out, forms, which are preserved in archaeological matter, are the product of a "vibration"—we would say of a tension field in time. That is why they evolve. Their morphology changes ever so slightly, much in the same way the anatomical features or organs of living creatures assume new forms. But, in every case, and no matter how spectacular the transformations, the skeleton—the internal organization of the skeleton—remains the same. This fundamental structure is what might be called the typological or iconological structure of material creations or representations. The study of objects as they are transformed over time is a matter of fleshing out this paleontology of forms.

This in no way means that we must abandon the basic historicist perspective of archaeology whose aim is to reproduce the material reality of the past. The failure to see artifacts as "memory-objects" has tended to push the discipline into an ancient approach that has been marginalized by historians. Post-war archaeology has proven remarkably incapable of incorporating the at times radical changes that have occurred in the field of history. In particular, it was largely impervious to those affected by the so-called New History of the 1960s and 1970s. Yet, one would have thought that archaeology would have been especially drawn to the way discrete, event-based historical time could be expanded to the multiple lengths of time in the "long-term history" that Fernand Braudel introduced at the end of the 1940s.[16] Similarly, when Emmanuel Le Roy Ladurie delved into the "micro-history" of a *langue d'oc* village between the end of the 13th and the beginning of the 14th centuries, or, conversely, when he turned to macro-history and studied the impact of climate change on medieval and modern societies, he was studying, over a variety of scales ranging from sites to landscapes to environments, time–space frameworks that are the common fare of archaeology.[17] Here again, one would have thought that this would bring archaeologists to re-examine their practices and reflect on the way history conceived of different temporal and spatial scales related to the traditional form of history based on "great events"; but it never happened, in spite of the appeals of historians.[18] Archaeologists proved incapable of answering the call. They had believed that artifacts explicitly bore witness to the reality of ancient times for so long that, paradoxically, they were incapable of seeing them as objects of history per se, as objects allowing us to re-examine history, or better yet, to challenge it.

The point then is not to reject history, but rather to assign it its true place in archaeology by making the discipline the "science of the past" that it has always sought to be. This is a theoretical matter; archaeology must define its objective: not the past, but what one might call that which is subject to the past. Whereas history seeks to establish what happened to people, archaeology explains what happened to things, in other words, to the material world of human communities. Operating over a variety of temporal and spatial ranges, archaeology studies how these things and beings have "absorbed" past events, both how they came to evolve on the basis of their situation and their inherited past, and how they contributed to these evolutionary processes.

Hence, archaeology is not the simple material illustration of history. It is rather a privileged field of observation that enriches history with the full range of individual trajectories of subjects from the material world. Viewed thus, it opens a series of unusually rich areas of study that allow us to observe not only how a single historical event can translate into beings and things in various situations, but how their interaction can generate further change that can, in turn, generate historical events.

The Cycles of Memory

Archaeology allows us to explore the processes at work in the formation of artifacts. These processes determine not only how these "archives of the past" come down to us, but how they were discovered and, beyond that, preserved. There exists, in particular, a *material memory cycle* over the course of which artifacts are altered, destroyed, buried, and perhaps (re) discovered, and then preserved as objects bearing witness to the past, and they may then be destroyed and "forgotten" all over again. One could cite a number of objects that have gone through several cycles of identification and disappearance: Greek statues, for example, that were copied by the Romans, then disappeared during the Middle Ages only to be rediscovered during the Renaissance and eventually reinterpreted in contemporary times as objects bearing witness to the history of civilization. In truth, all of material culture goes through such cycles, but the triviality of most things generally prevents us from discerning the pattern.

In the present, or more specifically, at the present, the mass of archaeological matter is largely composed of recent items whose numbers rapidly decrease as we move back in time. These are objects generally preserved above ground. They are part of the material we use every day. The oldest pieces may not be immediately visible to us, but they haven't disappeared. Through archaeology, we know that they are still a part of our present, but that they are not immediately accessible because they are buried or covered by other vestiges.

Moreover, the items that make up the material mass include not only functioning objects and structures whose historical identity goes unrecognized, but items that are no longer being used but that have been preserved as witnesses to the past: monuments and sites we refer to as "historic," and also archives, documents, and collections of objects and

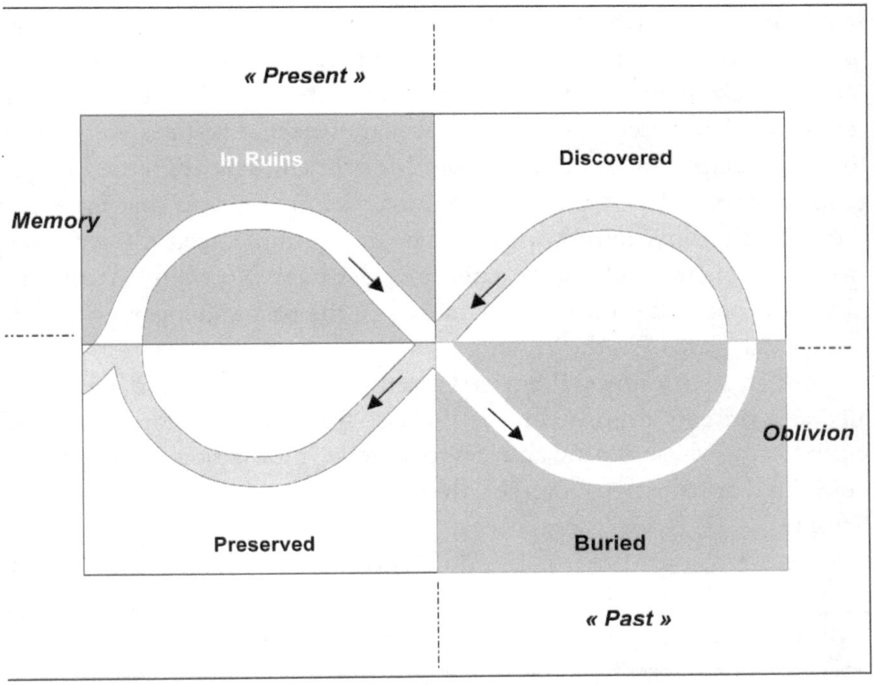

Diagram of the cycle of material memory. In white: the realm of the visible; in dark gray: the realm of the invisible. The light-gray circuit: the visible trajectory of vestiges; the white circuit: the hidden trajectory of vestiges. Things that are used end up ruined and buried. Things can be discovered only because they have been buried, which is to say, hidden. Things can be preserved only because they have been discovered. Preserved things, too, are destined to end up ruined and buried.

even artworks that are housed in libraries and museums. Here, too, the greater part of these material items are recent, and their numbers diminish rapidly as we move back in time. For example, there are far greater numbers of 19th- and 20th-century objects, documents, and structures than there are from the Middle Ages or, obviously, Antiquity. The chronological distinctions between items held in the present—either objects in use or those preserved for their historical value—have a direct impact on the conditions under which they are preserved, with recent items having a far greater probability of being destroyed or broken. We know, all too well, unfortunately, that fire and war strike foremost the material culture of the present.

When such is their fate, what remains is generally discarded or buried, for it is no longer functional and/or can no longer bear witness to the past.

It becomes trash, which is to say, a vestige adding to the considerable mass of remains from "the past," or more precisely, from a time that has "ended." Like present-day items—whose historical value we either fail to see or which we recognize as archival matter—these vestiges are subject to two principal forms of deposition. Most often, they are buried in the ground as secondary deposits. Like most material remains that have lost their functionality, most buried artifacts have simply been thrown away (what we generally refer to as garbage or detritus). No willful attempt to identify them as documents or witnesses to the past accompanies the act of depositing them in the ground.

But at certain times, the tiniest fraction of them is buried intentionally, as a primary deposit. This is the case for tombs and for all sorts of buried items whose general aim it is to preserve information identifying an objectified or "conscious" present, the prime example being the Auschwitz Scrolls.

Here too, once buried, the difference between matter that has or hasn't been recognized to be of archival value fades. Once things disappear beneath the surface, they definitively leave the domain of the objectified or "conscious" present and become a part of the forgotten, "unconscious" past. We can put it another way: present-day matter belongs to an accessible history of which people can give objective accounts, whereas matter from the past belongs to a "prehistory" to which we do not have access in that, because it lies below ground, it cannot be understood through first-hand accounts. Both categories face similarly probable ends. Just as destruction awaits elements of the "conscious" present, so discovery awaits those of the "unconscious" past or of the "forgotten" present. But as opposed to elements of the conscious present—which are by and large drawn from the recent past—those of the unconscious past are basically drawn from distant times. All archaeologists working on digs know from experience that, should they find any artifacts at all, there is a very good chance that they will be relatively ancient. Systematic digs carried out over the last thirty years have even revealed that, statistically, artifacts from proto-historical periods and from ancient Rome are far more likely to be discovered on land development projects outside of cities.

This particular aspect of buried material memory has led archaeologists working in the field to concentrate on ancient vestiges rather than on recent items, even though they essentially belong to the same kind of archaeological remains. When we unearth artifacts and recognize them as such, they are included in the mass of material elements from the past and the "conscious" present. Collections are preserved in repositories, exhibited in museums, and even examined in laboratories for the information they can yield about the past. Because they are thrown in with matter from the conscious present, they face the same threat of destruction, which would return them once more to oblivion. Recent history has shown that museums and art and antique collections are often among the first victims of war and revolution.

It is now easier to see the fundamentally paradoxical nature of material vestiges that the conventional opposition between present and past divides into separate categories. As we have stated many times over, the present, understood as the sum of present-day material elements, is largely composed of vestiges of the past and, conversely, the past, understood as the sum of ancient material elements, continues to have a direct effect upon the present. The conventional distinction between past and present has effectively lost all meaning. What matters is rather the contextual difference between material elements that have been preserved and those that have been buried, between what has been identified and what has been forgotten, between, ultimately, the conscious and the unconscious. Matter from both present and past is either preserved or destroyed in the here-and-now, or buried in or unearthed from the past.

The archival situation of this material memory is paradoxical, not only because it has no truck with the divisions of conventional time, but also because it carries within it its own negation. With regard to the present, what we preserve as archival matter we can preserve as such only because of the mass destruction of the witnesses to present-day "history," which grants these relics, that are frequently chance finds, document status. Similarly, with regard to the past, or, if you will, to the "prehistory" of today's present, we can unearth vestiges only because of the massive disappearance of traces of the present. Vestiges are visible to us as such only because of the great mass of matter that is not present underground. This is why archives and vestiges are irrevocably filled with gaps; the materiality of the present to which these documents bear witness is inextricably preserved and destroyed, objectified and forgotten. This also explains why archives, or vestiges, are at once accessible and inaccessible. They are as much witnesses that have miraculously escaped destruction and oblivion as pure inventions of the circumstances of their discovery and the way they are exhibited. Historians and archaeologists invent the objects they study as much as they discover them. Above all, as Derrida pointed out, "repressed or suppressed . . . the truth is spectral, and this part of its truth cannot be reduced by explanation."[19]

Opting for the Little Thing

There remains a mystery to be solved. If this is really so, how is it, when all the processes of reproduction seem dominated by chance and uncertainty, that the data, which we know to be the remains of fragmented discards, can emerge so solidly ordered? It is a question that haunts all disciplines that deal with matter and time, foremost among them, obviously, physics. The somewhat classical response has it that a fundamental order lies at the heart of matter, and that this order is mathematical in nature. This conventional interpretation is part of the late 19th-century positivist tradition. It underlies most of today's mathematical approaches to archaeological data that look for patterns while eliminating the "noise." It is not the kind of response that can offer a cogent explanation of the problem matter poses. In fact, the order lies in the noise. The "noise" in the data—which conventional, statistical approaches specifically seek to remove—constitutes an order that can never be grasped in its entirety.

All of which means that there exists another possibility whose implications need to be explored more carefully, namely, that the order could create itself, as a sort of morphological property of matter in transformation. No longer would archaeologists need to look to the heart of social structures or cultural manifestations; order would be spontaneously generated as soon as something is reproduced and has amassed memory. This is precisely the process from which fractal shapes, that are so similar to forms in "real" nature, evolve, a process of constant reiteration that creates them and makes them grow. And shapes emerge and develop precisely because, every step of the way, each reiteration is at once undetermined yet dependent upon the previous history of the system as a whole. The reproduction of objects, the rebuilding of structures, and the accumulation of archaeological strata generate an order that we do not know how to see. What we call style, or cultural identity, can clearly be approached as one of these constructs, structured over time by the effects of reiteration. There are surely many others as well, but archaeological time is fundamentally derived from this kind of structure. If that be the case, then Miró was right: Even the humblest, simplest things are indicative of an extraordinary process, and an immense field of research lies open before us.

In effect, we are just getting started. Faced with a world like ours that has known devastation, we, like Adorno, probably have to "opt for the little thing,"[20] choose the insignificant, in other words, as the poet René Char said, favor the precarious and the rudimentary. As Walter Benjamin noted, when the exceptional becomes the rule,[21] when daily life in the post-industrial world becomes that "strange state in which the line separating war and peace tends to disappear, and yet which brings with it the blossoming of industrial activities, the arts, the sciences, metropolises, in short, the signs of a fully developed society,"[22] we have to understand that, if we have effectively reached this point, our conventional conception of the movement of history is "untenable." We have grown accustomed to living with the constant threat of, if not total destruction, at least massive destruction. We believe that we are relatively protected, without realizing that this illusion of collective comfort has had as its price the devaluation of what is human. Resisting this illusion means refusing to relinquish our individual portion of humanity by positing history as the locus of our conscience. It means regarding all the little things in the world as at least as important as the big ones, and receiving them, in their fragile nowness, as a marvelous present of lfe.

Notes

1. Giuseppe Penone, *The Structure of Time* (Annecy: DAO, 1993), p. 7.

2. Henri Bergson, *L'Évolution créatrice* (Paris: Presses universitaires de France, 1941), p. 16.

3. When he was already well on in years, Hokusai said: "When I was around fifty I did thousands of works, but nothing I did before I was seventy counts. It was when I was around seventy-three that I began to understand the true nature [of things] . . . . When I reach one hundred, everything in my work—points, lines—will be alive."

4. Joan Miró to J.-F. Rafols, Montroig, August 11, 1918.

5. Georges Pérec, *Species of Spaces and other Pieces,* trans. by John Sturrock. (London: Penguin Books, 1997), pp. 24–25.

6. "The history of art," Winckelmann said, "should inform us about the origin, growth, change, and fall of art, together with the various styles of peoples, periods, and artists." Johann Joachim Winckelmann, *History of the Art of Antiquity*, trans. Harry Francis Mallgrave (Los Angeles: Getty Publications, 2006), p. 71.

7. For example, the ethnologist Philippe Descola has demonstrated that the basic distinction between nature and culture is merely one specifically modern, Western way of conceiving the world, simply a variation of the four basic modes of identifying beings and things that human societies develop. Philippe Descola, *Par-delà nature et culture* (Paris: Gallimard, 2005).

8. In his study of the way in which anthropology depicts time, Johannes Fabian pointed out that "anthropology emerged and established itself as an allochronic discourse: it is a science of other men in another Time." Johannes Fabian, *Time and the Other* (New York: Columbia University Press, 1983), p. 143. Nicholas Thomas also noted the "out of time" dimension of anthropology. Its object, he said, "was and is essentially a social or cultural system or structure out of time." Nicholas Thomas, *Out of Time* (Cambridge: Cambridge University Press, 1989), p. 120.

9. This approach, driven in particular by structuralism, was criticized in the 1960s by Paul Ricœur, especially in *Le Conflit des interprétations. Essai d'herméneutique,* vol. 1 (Paris: Le Seuil, 1969).

10. Gérard Chouquer, "*Crise et recomposition des objets: les enjeux de l'archéologie,*" in Gérard Chouquer, "Objets en crise, objecs recomposés. Transmissions et transformations des espaces historiques. Enjeux et contours de l'archéogéographie," in *Études rurales,* no. 167–8, 2004, pp. 13–32.

11. François Hartog, *Régimes d'historicité. Présentisme et expériences du temps* (Paris: Le Seuil, 2003).

12. Paul Ricœur, *De l'Interprétation. Essai sur Freud* (Paris: Le Seuil, 1965); Paul Ricœur, *Temps et récit,* vol. 1, *L'Intrigue et le temps historique* (Paris: Le Seuil, 1983).

13. Jorge Luis Borges, *Ficciones*, trans. Emecé Editores, Buenos Aires (New York: Grove Press, 1962), pp. 112–3.

14. Jean-Luc Godard, Interview in *Libération*, May 15–16 (2004), p. 49.

15. Heraclitus, *Fragments. Citations et témoignages* (Paris: Flammarion, 2002), pp. 66–67.

16. Fernand Braudel, *La Méditerranée et le monde méditérranéen à l'époque de Philippe II* (Paris: Armand Colin, 1949).

17. Emmanuel Le Roy Ladurie, *Moutaillou, village occitan de 1294 à 1324* (Paris: Gallimard, 1975); Emmanuel Le Roy Ladurie, *Histoire du climat depuis l'an Mil* (Paris: Flammarion, 1967).

18. Jean-Marie Pesez, "Histoire de la culture matérielle," in J. Le Goff, *La Nouvelle Histoire* (Paris: Complexe, 1988).

19. Jacques Derrida, *Archive Fever, A Freudian Impression*, trans. Eric Prenowitz (Chicago: The University of Chicago Press, 1996), p. 87.

20. Miguel Abensour, postface to Theodor W. Adorno, *Minima Moralia. Réflexions sur la vie mutilée*, trans. Éliane Kaufholz and Jean-René Ladmiral (Paris: Payot, 2003).

21. Benjamin, *Selected Writings*, p. 392.

22. Abensour, in Adorno, *Minima*, p. 335.

BIBLIOGRAPHY

Abensour, Miguel. Postface to Theodor W. Adorno, *Minima Moralia. Réflexions sur la vie mutilée*, translated by Éliane Kaufholz and Jean-René Ladmiral. Paris: Payot, 2003, pp. 335–54.

Ackerman, Michael. *End Time City*. Paris: Nathan-Delpire, 1999.

Ackerman, Michael. *Fiction*. Paris: Delpire, 2001.

Adam, Frédéric. *Alain-Fournier et ses compagnons d'armes. Une archéologie de la Grande Guerre*. Metz: Serpenoise, 2006.

Agache, Roger and Bruno Bréart. *Atlas d'archéologie aérienne de Picardie*. Amiens: Société des Antiquaires de Picardie, 1975.

Agacinski, Sylviane. *Le Passeur de temps. Modernité et nostalgie*. Paris: Le Seuil, 2000.

Agamben, Giorgio. *Remnants of Auschwitz*, translated by Daniel Heller-Roazen. New York: Zone Books, 1999.

Agamben, Giorgio. *Image et mémoire. Écrits sur l'image, la danse et le cinéma*. Paris: Desclée de Brouwer, 2004.

Alcock, Susan E. *Archaeologies of the Greek Past. Landscape, Monuments and Memories*. Cambridge: Cambridge University Press, 2002.

Alexievitch, Svetlana. *Tchernobylskaïa molita*. Moscow: Ostojié, 1997.

Alexievitch, Svetlana. *Voices from Chernobyl*, translated by Keith Gessen. London: Dalkey Archive Press, 2005.

Anonymous. Sentence dated January 3, 1945, in *Revue d'histoire de la Shoah*, 171 (2001), pp. 162–7.

Arendt, Hannah. *Between Past and Future; Eight Exercises in Political Thought*. New York: Viking, 1972.

BIBLIOGRAPHY

Bailey, Geoff. "Concepts, Time-scales and Explanations in Economic Prehistory," in A. N. Sheridan and Geoff Bailey, *Economic Archaeology*. Oxford, British Archaeological Reports, International Series (1981), pp. 97–117.

Bailey, Geoff. *Palimpsests, Time-scales, and the Durational Present: Origins and Consequences of a Time-Perspective Research Agenda*. University of Newcastle, unpublished manuscript, 2003.

Barthes, Roland. *Camera Lucida*, translated by Richard Howard. New York: Hill and Wang, 1982.

Beck, Ulrich. *Risk Society: Towards a New Modernity*, translated by Mark Ritter. London: Sage Publications, 1992.

Benjamin, Walter. *Œuvres*, translated by Rainer Rochliz and Pierre Rusch. Paris: Gallimard, 2000.

Benjamin, Walter. *Selected Writings*, translated by Edmund Jephcott and others. Cambridge, MA: Harvard University Press, 2003.

Benjamin, Walter. *Sens unique. Une enfance berlinoise. Paysages urbains*. Paris: Maurice Nadeau, 1994.

Benjamin, Walter. *The Arcades Project*, trans. Howard Eiland and Kevin McLaughlin. Cambridge, MA: The Belknap Press of Harvard University Press, 1999.

Bergson, Henri. *L'Évolution créatrice*. Paris: Presses universitaires de France, 1941.

Bergson, Henri. *Matières et mémoire*. Paris: Presses universitaires de France, 1896.

Berque, Augustin. *Écoumène. Introduction à l'étude des milieux humains*. Paris: Belin, 2000.

Bertrand, Alexandre. "Les Tombelles d'Auvenay (Côte d'Or)," in *Revue archéologique* IV (1861), pp. 1–11.

Bertrand, Élie. *Dictionnaire universel des fossiles propres et des fossiles accidentels*. Avignon: Louis Chambeau, 1763.

Biel, Jörg. *Der Keltenfürst von Hochdorf*. Stuttgart, Konrad Theiss, 1985.

Binford, Lewis. "Behavioral Archaeology and the 'Pompeii Premise,'" *Journal of Anthropological Research* 37 (1981), pp. 195–208.

Binford, Lewis. *In Pursuit of the Past*. London: Thames & Hudson, 1983.

Binswanger, Ludwig and Aby Warburg. *La Guérison infinie. Histoire clinique d'Aby Warburg*, translated by Maël Renouard and Martin Rueff. Paris: Payot et Rivages, 2007.

Blanchot, Maurice. *The Writing of the Disaster*, translated by Ann Smock. Lincoln: University of Nebraska Press, 1986.

Bleicher, Jules and Jules Beaupré. *Guide pour des recherches archéologiques (époques préhistorique, gallo-romaine et mérovingienne) dans l'Est de la France. Belfort, Doubs, Haute-Saone, Meurthe-et-Moselle, Meuse, Vosges*. Nancy: Crépin-Leblond, 1896.

Bonnichsen, Robson. "Millie's Camp: An Experiment in Archaeology," *World Archaeology*, 4, 3 (1973), pp. 277–91.

Borges, Jorge Luis. "Kafka and his Precursors," translated by James E. Irby. Online: http://sillysoft.net/forums/viewtopic.php?t=1214.

Borges, Jorge Luis. *Ficciones*, translated by Emecé Editores, Buenos Aires. New York: Grove Press, 1962.

Botting, Fred and J.K. Davies. "Freud's Library and an Appendix of Texts Related to Antiquities." In Lynn Gamwell and Richard Wells, *Sigmund Freud and Art. His Personal Collection of Antiquities*. London and New York: Freud Museum and State University of New York, 1989, pp. 184–92.

Bowden, Mark. *The Life and Archaeological Work of Lieutenant-General Augustus Henry Lane Fox Pitt Rivers, DCL, FRS, FSA*. Cambridge: Cambridge University Press, 1991.

Bradley, Richard. *The Past in Prehistoric Societies*. London, New York: Routledge, 2002.

Braudel, Fernand. *Écrits sur l'histoire*. Paris: Flammarion, 1969.

Braudel, Fernand. *La Méditerranée et le monde méditérranéen à l'époque de Philippe II*. Paris: Armand Colin, 1949.

Brézillon, Michel. "La Fouille" in *Dictionnaire de la Préhistoire*. Paris: Larousse, 1969.

Buffetaut, Éric. *Histoire de la Paléontologie*. Paris: Presses universitaires de France, 1998.

Carter, Howard and A.C. Mace. *The Tomb of Tut-Ankh-Amen. Discovered by the late Earl of Carnavaron and Howard Carter*. London: Cassel, 1923–1933.

Castor, Markus. "Caylus et le cercle artistique parisien," in Aghion, Irène. *Catalogue of the exhibit of the Bibliothèque Nationale's Department of coins, medallions and antiquities*. Paris: Institut national de l'histoire de l'art, 2002, pp. 37–43.

Caylus, Anne Claude Philippe. *Recueil d'antiquités égyptiennes, grecques, étrusques et romaines*, 7 volumes. Paris: Desant & Saillant, 1752–1767.

Cendrars, Blaise. *L'Homme foudroyé. Le Sans-Nom*. Paris: Denoël, 2002.

Certeau, Michel de. *Histoire et psychanalyse. Entre science et fiction*. Paris: Gallimard, 1987.

Certeau, Michel de. *L'Écriture de l'histoire*. Paris: Gallimard, 1975.

Chouquer, Gérard. "Crise et recomposition des objets: les enjeux de l'archéologie," in Gérard Chouquer, "Objets en crise, objets recomposés. Transmissions et transformations des espaces historiques. Enjeux et contours de l'archéogéographie," in *Études rurales*, no. 167–8, 2004, pp. 13–32.

Cicero. *Academica*, in *Cicero, De Natura Deorum, Academica*, translated by H. Rackham. London: William Heinemann, 1933.

Clarke, David L. *Analytical Archaeology*. London: Methuen, 1978.

Cohen, Claudine and Jean-Jacques Hublin. *Boucher de Perthes. Les Origines romantiques de la préhistoire*. Paris: Belin, 1989.

Darwin, Charles, *Autobiography,* edited by Nora Barlow. London: Collins, 1958. Online: http://darwin-online.org.uk/content/frameset?itemID=F1497&viewtype=text&pageseq=1 (p. 119).

Darwin, Charles. *On the Origin of Species by Means of Natural Selection, or The Preservation of Favoured Races in the Struggle for Life*. London: John Murray, 1859.

Darwin, Charles. *The Correspondence of Charles Darwin*, edited by F. Buckhardt and S. Smith. Cambridge, Cambridge University Press, 1987.

Darwin, Charles. *The Formation of Vegetable Mould through the Action of Worms, with Observations on their Habits.* London: John Murray, 1881.

Darwin, Charles. *The Variation of Animals and Plants under Domestication.* London: John Murray, 1868.

Déchelette, Joseph. *Manuel d'archéologie préhistorique, celtique et gallo-romaine, I, Archéologie préhistorique*. Paris: Picard, 1908.

Déchelette, Joseph. *Manuel d'archéologie préhistorique, celtique et gallo-romaine, II, Archéologie celtique ou protohistorique, Âge du Bronze.* Paris: Picard, 1910.

Déchelette, Joseph. *Manuel d'archéologie préhistorique, celtique et gallo-romaine, II, Archéologie celtique ou protohistorique 2, Premier Âge du Fer ou époque de Hallstatt.* Paris: Picard, 1913.

Déchelette, Joseph. *Manuel d'archéologie préhistorique, celtique et gallo-romaine, II, Archéologie celtique ou protohistorique 3, Second Âge du Fer ou époque de la Tène.* Paris: Picard, 1914.

Deleuze, Gilles. *Nietzsche and Philosophy,* trans. by Hugh Tomlinson. New York: Columbia University Press, 1983.

Demoule, Jean-Paul, et al. *Guide des méthodes de l'archéologie.* Paris: La Découverte, 2002.

Derrida, Jacques. *Archive Fever, A Freudian Impression,* translated by Eric Prenowitz. Chicago: University of Chicago Press, 1996.

Descola, Philippe. *Par-delà nature et culture*. Paris: Gallimard, 2005.

Didi-Huberman, Georges. *Devant le temps. Histoire de l'art et anachronisme des images*. Paris: Minuit, 2000.

Didi-Huberman, Georges. *Génie du non-lieu. Air, poussière, empreinte, hantise.* Paris: Minuit, 2001.

Didi-Huberman, Georges. *L'Image survivante. Histoire de l'art et temps des fantômes selon Aby Warburg*. Paris: Minuit, 2002.

Didi-Huberman, Georges. *Ninfa Moderna. Essai sur le drapé tombé.* Paris: Gallimard, 2002.

Dörpfeld, Wilhelm. *Troja und Ilion. Ergebnisse der Ausgrabungen in der vorhistorischen und historischen Schlichten von Ilion, 1870–1894.* Athens: Beck & Barth, 1902.
Droop, P. J. *Archaeological Excavation.* Cambridge: Cambridge University Press, 1915.
Dufour-El Maleh, Marie-Cécile. *La Nuit sauvée. Walter Benjamin et la pensée de l'Histoire.* Bruxelles: Ousia, 1993.
Enzenberger, H.M. *L'Europe en ruines*, translated by Bernard Kreiss. Arles: Actes Sud, 1995.
Fabian, Johannes. *Time and the Other.* New York: Columbia University Press, 1983.
Farmer, Sarah. *Oradour. Arrêt sur mémoire*, trans. by Pierre Guglielmina. Paris: Calmann-Lévy, 1994.
Febvre, Lucien. *Combats pour l'histoire.* Paris: Armand Colin, 1953.
Focillon, Henri. *The Life of Forms in Art.* New Haven, CT: Yale University Press, 1942.
Foucault, Michel. "Il faut défendre la société." *Cours au Collège de France 1976–1977.* Paris: Le Seuil-Gallimard, 1997.
Foucault, Michel. *The Order of Things. An Archaeology of the Human Sciences.* New York: Pantheon, 1970.
Fouillée, Augustine, pseudonym G. Bruno. *Le Tour de la France par deux enfants. Devoir et patrie.* Paris: Eugène Belin, 1877.
Fradin, Émile. *Glozel et ma vie.* Paris: Laffont, 1979.
Freud, Sigmund. *Études sur l'hystérie*, translated by Anne Berman. Paris: Presses universitaires de France, 1956.
Freud, Sigmund. *Le Délire et les rêves dans la Gradiva de W. Jensen*, translated by Paule Arbex and Rose-Marie Zeitlin. Paris: Gallimard, 1986.
Freud, Sigmund. *Letters of Sigmund Freud,* edited by Ernst Freud, translated by Elaine and William Robson-Scott. New York: Basic Books, 1960.
Freud, Sigmund. *Psycholanalysis and Faith. The Letters of Sigmund Freud and Oskar Pfister*, edited by Heinrich Meng and Ernst L. Freud, translated by Eric Mosbacher. New York: Basic Books, 1963.
Freud, Sigmund. *The Complete Letters of Sigmund Freud to Wilhelm Fliess, 1877–1904*, translated by Jeffrey Moussaieff Masson. Cambridge, MA: The Belknap Press of Harvard University Press, 1985.
Fukuyama, Francis. *The End of History and the Last Man.* New York: Free Press, 1992.
Galles, René. "*Rapport à M. le préfet du Morbihan sur les fouilles du Mont-Saint-Michel en Carnac faites en septembre 1862.*" Vannes: Imprimerie Galles, 1862.
Giddens, Anthony. *Consequences of Modernity.* Cambridge: Polity Press, 1990.

Godard, Jean-Luc. Interview in *Libération*, May 15–16, 2004, p. 49.

Gombrich, Ernst Hans. *Aby Warburg. An Intellectual Biography*. London: University of London, Warburg Institute, 1970.

Gradowski, Zalmen, "Megilot Auschwitz. Les Rouleaux d'Auschwitz (I)," in *Revue d'histoire de la Shoah*, 171 (2001), pp. 18–126.

Gräslund, Bo. *The Birth of Prehistoric Chronology. Dating Methods and Dating Systems in Nineteenth Century Scandinavian Archaeology*. Cambridge: Cambridge University Press, 1987.

Green, André. *Diachrony in Psychoanalysis*, translated by Andrew Weller. London: Free Association Books, 2003.

Grenier, Albert. *Manuel d'archéologie gallo-romaine, II, L'Archéologie du sol (vol. 1, Les Routes; vol. 2, L'Occupation du sol et index des deux volumes)*. Paris: Picard, 1934.

Grenier, Albert. *Manuel d'archéologie gallo-romaine, III, L'Architecture (vol. 1, Notions préliminaires. L'urbanisme. Les monuments de la vie politique: capitole, forum, temple, basilique; vol. 2, Ludi et Circenses: théâtre, amphithéâtre, cirque)*. Paris: Picard, 1958.

Grenier, Albert. *Manuel d'archéologie gallo-romaine, I, Généralités et travaux militaires*. Paris: Picard, 1931.

Grenier, Albert. *Manuel d'archéologie gallo-romaine, IV, Les Monuments des eaux (vol. 1, Aqueducs, thermes; vol. 2, Villes d'eau, sanctuaires de l'eau*. Paris: Picard, 1960.

Hartog, François. *Régimes d'historicité. Présentisme et expériences du temps*. Paris: Le Seuil, 2003.

Hatt, Jean-Jacques. "Pour connaître le passé des villes gallo-romaines. La méthode stratigraphique des fouilles (résultats obtenus à Strasbourg)," in *Revue archéologique de l'Est*, II (1951), pp. 44–51.

Heraclitus. *Fragments. Citations et témoignages*. Paris: Flammarion, 2002.

Hildebrand, Hans. *De förhistoriska folken i Europa. En handbok i jämförande fornkunskap*. Stockholm, 1873–1880.

Hildebrand, Hans. *Den vetenskapliga fornsforkningen, hennes uppgift, behof och rätt*. Stockholm, 1873.

Hobsbawm, Eric. *L'Âge des extrêmes. Histoire du court XXe siècle*. Paris: Complexe, 1994.

Hodder, Ian. "Éthnoarchéologie: une approche conceptuelle" in *Les Nouvelles de l'Archéologie* (1980), pp. 24–30.

Hodder, Ian. *Symbols in Action*. Cambridge: Cambridge University Press, 1981.

Hubert, Henri. "La Collection Moreau au Musée de Saint-Germain," *Revue archéologique*, XLI (1902), pp. 167–208.

Hubert, Henri. *Les Celtes et l'expansion celtique depuis l'époque de la Tène et la civilisation celtique.* Paris: Albin Michel, 1950.

Hubert, Henri. *Les Germains.* Paris: Albin Michel, 1952.

Isambert, François-André. "Henri Hubert et la sociologie du temps," *Revue française de sociologie*, XX (1979).

Jullian, Camille. "Les Fouilles de Glozel," in *Comptes rendus de l'Académie des Inscriptions et Belles Lettres.* Paris: Picard, 1926, pp. 257–9.

Jung, C. "Morphogenèse, fonctions et évolution de la centuriation B d'Orange et essai de restitution diachronique des paléopaysages du Tricastin (Drôme-Vaucluse)" (Tours: Université de Tours, unpublished doctoral thesis, 2 volumes).

Klein, Robert. *La Forme et l'intelligible.* Paris: Gallimard, 1970.

Kracauer, Siegfried. *History, The Last Things before the Last.* New York: Oxford University Press, 1969.

Kubler, George. *The Shape of Time.* New Haven, CT: Yale University Press, 1962.

Lacan, Jacques. Interview with Gilles Lapouge in *Le Figaro littéraire*, no. 1076 (1966), p. 2.

Laming-Emperaire, Annette. *Origines de l'archéologie préhistorique en France. Des superstitions médiévales à la découverte de l'homme fossile.* Paris: Picard, 1964.

Lanzmann, Claude. *Shoah, An Oral History of the Holocaust.* New York: Pantheon, 1985.

Latour, Bruno. *La Science en action. Introduction à la sociologie des sciences.* Paris: Gallimard, 1995.

Latour, Bruno. *Nous n'avons jamais été modernes. Essai d'anthropologie symétrique.* Paris: La Découverte, 1989.

Le Goff, Jacques. *La Nouvelle Histoire.* Paris: Complexe, 1988.

Le Roy Ladurie, Emmanuel. *Histoire du climat depuis l'an Mil.* Paris: Flammarion, 1967.

Le Roy Ladurie, Emmanuel. *Moutaillou, village occitan de 1294 à 1324.* Paris: Gallimard, 1975.

Ledig, Gert. *Payback,* translated by Shaun Whiteside. London: Granta Books, 2003.

Legendre, Jean-Pierre, Laurent Olivier, and Bernadette Schnitzler. *L'Archéologie nazie à l'Ouest du Reich.* Paris: Infolio, 2007.

Leroi-Gourhan, André and Michel Brézillon. *Fouilles de Pincevent. Essai d'analyse ethnographique d'un habitat magdalénien.* Paris: CNRS, 1972.

Leroi-Gourhan, André. *Les Fouilles préhistoriques (techniques et méthodes).* Paris: A. et J. Picard, 1950.

Levi, Primo. *If This Is a Man.* New York: Collier Books, 1961.

Lowenthal, David. *The Past Is a Foreign Country.* Cambridge: Cambridge University Press, 1985.

Löwy, Michael. *Walter Benjamin. Avertissement d'incendie. Une lecture des thèses "Sur le concept d'histoire."* Paris: Presses universitaires de France, 2001.

Lyell, Charles. *Principles of Geology, Being an attempt to explain the former changes of the Earth's surface by reference to causes now in operation.* London: John Murray, 1830.

Lyotard, Jean-François. *Discours, Figures.* Paris: Klincksieck, 1971.

Malthus, Thomas. *An Essay on the Principle of Population.* Oxford: Oxford University Press, 1993.

Mauss, Marcel. "Avertissement" to Hubert, Henri. *Les Celtes et l'expansion celtique depuis l'époque de la Tène et la civilisation celtique.* Paris: Albin Michel, 1932, xxi–xxvi.

Michaud, Philippe-Alain. *Aby Warburg and the Image in Motion,* translated by Sophie Hawkes. New York: Zone Books, 2004.

Michaud, Philippe-Alain. *Aby Warburg et l'image en mouvement, suivi de Aby Warburg, souvenirs d'un voyage en pays Pueblo, 1923, et Projet de voyage en Amérique.* Paris: Macula, 1998.

Montelius, Oscar. "Den Förhistoriske fornforskarens metod och material," in *Antiqvarisk Tidskrift för Sverige,* 8 (1884), pp. 1–28.

Montelius, Oscar. "Typologien eller utvecklingsläran tillämpad pa det menskliga arbetet," in *Svenska Fornminnföreningens Tidskrift,* 3 (1899), pp. 237–68.

Montfaucon, Bernard de. *Antiquity explained, and represented in sculptures, By the Learned Father Montfaucon,* translated by David Humphreys. London: J. Jonson and J. Watts, 1721.

Mortillet, Gabriel de. "Essai d'une classification des cavernes et des stations sous abri fondée sur les produits de l'industrie humaine," *Matériaux pour l'Histoire primitive et naturelle de l'Homme 5* (1869), pp. 172–9.

Mortillet, Gabriel de. *Le Préhistorique. Antiquité de l'Homme.* Paris: Reinwald, 1883.

Nietzsche, Friedrich. *Fragments posthumes sur l'éternel retour,* translated by Lionel Duvoy. Paris: Allia, 2003.

Nietzsche, Friedrich. *Œuvres philosophiques complètes,* translated by P. Rusch. Paris: Gallimard, 1990.

Nisard, Charles. *Correspondance inédite du comte de Caylus avec le P. Paciaudi, théatin* (1757–1765). Paris: Imprimerie nationale, 1877.

Norton Cru, Jean. *Du témoignage.* Paris: Jean-Jacques Pauvert, 1967.

Norton Cru, Jean. *Témoins. Essai d'analyse et de critiques des souvenirs de combattants édités en français, de 1915 à 1928.* Paris: Les Étincelles, 1929.

Öe, Kenzaburö. *Hiroshima Notes*, translated by David L. Swain and Toshi Yonezawa. New York: Grove Press, 1996.

Olivier, Laurent and Bruno Wirtz. "Recherches sur le temps archéologique: l'apport de l'archéologie du présent" in *Antiquités nationales*, 35 (2003), pp. 255–66.

Olivier, Laurent. "Henri Hubert, archéologue," in Brun, P. and Laurent Olivier. *Les Nouvelles de l'archéologie*, 79, pp. 9–14.

Olivier, Laurent. "Nouvelles recherches sur le tumulus à tombe à char de La Butte à Sainte-Colombe-sur-Seine (Côte-d'Or)," in *Antiquités nationales*, 32 (2000), pp. 97–115.

Olivier, Laurent. "The Hochdorf Princely Grave and the Question of the Nature of Archaeological Funerary Assemblages," in Tim Murray, *Time and Archaeology*. London, New York: Routledge, 1999, pp. 109–38.

Patrick, L. "Is There an Archaeological Record?" in *Advances in Archaeological Method and Theory*, 8 (1985).

Penone, Giuseppe. *La Structure du temps*. Annecy: DAO, 1993.

Pérec, Georges. *Ellis Island*, translated by Harry Matthews and Jessica Blatt. New York: The New Press, 1995.

Pérec, Georges. *Life, A User's Manual*, translated by David Bellos. Boston: David R. Godine, 1978.

Pérec, Georges. *Species of Spaces and other Pieces*, translated by John Sturrock. London: Penguin Books, 1997.

Pesez, Jean-Marie. "Histoire de la culture matérielle," in J. Le Goff, *La Nouvelle Histoire*. Paris: Complexe, 1988, pp. 191–227.

Pitt Rivers, Augustus. "Typological Museums" in *Journal of the Society of Arts*, no. 40, December 18, 1891, pp. 115–22.

Plato. *Hippias Major*, translated by Benjamin Jowett. Online: http://www.ellopos.net/elpenor/greek-texts/ancient-greece/plato/plato-hippias-major.asp?pg=8.

Pomian, Krzystzof. *Collectionneurs, amateurs et curieux. Paris-Venise XVIe-XVIIIe siècle*. Paris: Gallimard, 1987.

Proust, Françoise. *L'Histoire à contre-temps. Le Temps historique chez Walter Benjamin*. Paris: Éditions du Cerf, 1994.

Proust, Marcel. *Remembrance of Things Past*, translated by C. K. Scott Moncrieff and Terence Kilmartin. New York: Random House, 1981.

Reinach, Salomon. "Les Fouilles de Glozel," in *Comptes rendus de l'Académie des Inscriptions et Belles Lettres*. Paris: Picard, 1926, pp. 203–5, 207, 218–22; 1927, pp. 197–8, 201–3.

Reinach, Salomon. *Glozel, la découverte, la controverse, les enseignements*. Paris: Krie, 1928.

Renfrew, Colin. *Before Civilization. The Radiocarbon Revolution and Prehistoric Europe*. London: Cape, 1973.

Ricœur, Paul. *De l'Interprétation. Essai sur Freud*. Paris: Le Seuil, 1965.

Ricœur, Paul. *Le Conflit des Interprétations. Essai d'herméneutique*, vol. 1. Paris: Le Seuil, 1969.

Ricœur, Paul. *Temps et récit*, vol. 1, *L'Intrigue et le temps historique*. Paris: Le Seuil, 1983.

Rossi, Paolo. *The Dark Abyss of Time: The History of the Earth and the History of Nations from Hooke to Vico*. Chicago: University of Chicago Press, 1984.

Rossignol, Claude. "Rapport sur les fouilles faites sur les Chaumes d'Auvenay, à une demi-lieue d'Ivry," in *Mémoires de la Commission des Antiquités du Département de la Côte d'Or*, 2 (1842–1846), pp. 79–83.

Rossignol, Claude. *Découverte d'armes gauloises et romaines réunies, faites dans la plaine des Laumes, sous les murs d'Alise (Côte-d'Or)*, 19-page manuscript. Saint-Germain-en-Laye: Archives of the National Museum of Archaeology.

Saulcy, Félicien de. "Guerre des Helvètes. Première Campagne de César," in *Revue archéologique*, 2 (1860), pp. 165–86, 242–59, 313–44.

Schiffer, Michael. *Formation Processes of the Archaeological Record*. Albuquerque: University of New Mexico Press, 1987.

Schliemann, Heinrich. *Ilios, Stadt und Land der Trojaner. Forschungen und 57-Entdeckungen in der Troas und besonders auf der Baustelle Troja*. Leipzig: F.A. Brockhaus, 1881.

Schliemann, Heinrich. *Mykenae: Bericht über meine Forschungen und Endeckungen in Mykenae und Tiryns*. Leipzig: F.A. Brockhaus, 1878.

Schnapp, Alain. *L'Archéologie d'aujourd'hui*. Paris: Hachette, 1980.

Schnapp, Alain. *La Conquête du passé. Aux origines de l'archéologie*. Paris: Carré, 1993.

Scollar, Irwin et al. *Archaeological Prospecting and Remote Sensing*. Cambridge: Cambridge University Press, 1990.

Sebald, Winfried. *On the Natural History of Destruction*, translated by Anthea Bell. New York: Random House, 2003.

Severi, Carlo. "Warburg anthropologue, ou le déchiffrement d'une utopie. De la biologie des images à l'anthropologie de la mémoire," in *L'Homme, revue française d'anthropologie* (2003), 165, pp. 77–128.

Stolpe, Hjalmar, *Collected Essays on Ornamental Art*. Stockholm: Aftonbladets tryckeri, 1927.

Strover, Éric and Gilles Peress. *Les Tombes. Srebrenica et Vukovar*. Zurich: Scalo, 1998.

Swiebocka, Teresa. *Auschwitz, A History in Photographs*. Bloomington and Indianapolis: Indiana University Press, 1999.

Thomas, Nicholas. *Out of Time*. Cambridge: Cambridge University Press, 1989.

Thompson, Michael. *General Pitt Rivers: Evolution and Archaeology in the Nineteenth Century*. Bradford-on-Avon: Moonraker Press, 1977.

Tylor, Edward Burnett. *Primitive Culture*. London: John Murray, 1871.

Virilio, Paul. *L'Insécurité du territoire*. Paris: Galilée, 1993.

Warburg, Aby. *Le Rituel du Serpent. Art et anthropologie. Récit d'un voyage en pays pueblo*. Paris: Macula, 2003.

Warburg, Aby. *Schlangenritual. Ein Reisebericht*. Berlin: Klaus Wagenbach, 1988.

Wheeler, Mortimer. *Archaeology from the Earth*. Oxford: Oxford University Press, 1954.

Winckelmann, Johann Joachim. *Geschichte der Kunst des Altherthums*. Dresden: Walther, 1764.

Winckelmann, Johann Joachim. *History of the Art of Antiquity*, translated by Harry Francis Mallgrave. Los Angeles: Getty Publications, 2006.

ABOUT THE AUTHOR AND TRANSLATOR

Laurent Olivier is curator of the Department of Celtic and Gallic archaeology at the National Museum of Archaeology at Saint-Germain-en-Laye (France) and teaches the history of archaeology at the Ecole Pratique des Hautes Etudes, Paris. He graduated from the University of Cambridge (UK) and the University of Paris (Pantheon-Sorbonne), were he got his Ph.D. in archaeology and a doctorate thesis in prehistory and anthropology, as well as a Habilitation à Diriger des Recherches (HDR) degree.

Olivier is a specialist of Celtic and Gallic archaeology. He is coordinating an interdisciplinary research project devoted to the study of "proto-industrial" salt production in Lorraine and its long-term impact on social and ecological environment. A large part of his research is focused on the role played by archaeology in the building of collective identities. He is preparing a book about the ideological use of European archaeology made by Nazism.

Arthur Greenspan has been a professional translator for almost thirty years, publishing regularly in both French and English. His translations include a number of works by James M. Cain, Robert B. Parker, and Val McDermid and Édouard's Daladier's prison journals. He splits his time between Maine, where he teaches French at Colby College, specializing in twentieth-century literature and film, and Paris, where he leads the good life.

www.ingramcontent.com/pod-product-compliance
Lightning Source LLC
LaVergne TN
LVHW101447200926
848483LV00001B/72

* 9 7 8 0 7 5 9 1 2 0 4 5 7 *